# THE UNKNOWN AND THE UNKNOWABLE

## RAMAN SIVASHANKAR

ISBN 979-8-88959-622-6

# Dedication

*I dedicate this book to*

*My wife Durga,*

*My two daughters Rukmini, Aishwarya and son Madhavan,*

*Their spouses Arun, Pradeep and Sunita,*

*My grandchildren Arjun, Riya, Aaditya, Priyanka, Amolika and Areana.*

*Without their whole hearted understanding and support my effort could not have reached fruition.*

*To the latest addition to our family, my senior grandson Arjun's wife Molly.*

*May the tribe increase and prosper.*

# Contents

# Preface

*The Unknown and the Unknowable*

is the sequel to my earlier book *Our Universe An Unending Mystery*, published by Create Space, USA in 2017. In this book I had described the Physical, Spiritual and Occult Worlds. While doing so I had referred to *Knowledge Frontier Areas* on a variety of topics where our present knowledge is incomplete.

In my present book, I have expanded on some of these topics and added a few new ones like the *Human Life Form* and *Math Conundrums* to make the review more comprehensive.

The dichotomy between the Spiritual and Scientific Worlds that I had touched upon earlier is further researched and their interdependence highlighted. Einstein had stated - *Science without Religion is lame, Religion without Science is blind.* At a more mundane level we tend to take the Universe for granted.

Endless days and nights pass by, seasonal weather cycles repeat. While most people do not spare a thought for it, the curious are satisfied with the simplistic explanation that the events occur because of the the earth's rotation on a tilted axis and its revolution around the sun. The common tendency is to analyze events from a *what, where,* and *when* point of view rather than *why* and *how?*

Take for example the familiar sights of sunrise and sunset. In South India in the month of April sunrise is at about 6.00 AM over the eastern horizon and sunset at 6.00 PM in the west. If you ask a guy on the street *why* and *how* these phenomena occur, he would look puzzled and respond:

*Who knows? Never thought about it.*

This typical response from the public motivated me to write this book.

**Part I** explores *the Human Life Form i.e. Ourselves*. The human life cycle spanning conception, childbirth, adolescence, adulthood, old age, and death are so commonplace that the intricacy and mystery that surround them remain hidden. The intricate working of major organs and glands, features unique to humans like the use of *languages* for precise communication, *proprioception, human consciousness, free will* and the esoteric belief in the *third eye* are discussed. I have allocated some space for *cancer* - cause of the disease, its prevention, treatment and recent theories.

Frontline topics in the Physical World, like *time travel and wormholes, quasars and pulsars, weird aspects of particle physics, the stupendous vastness of inter galactic space* are probed.

Selected topics in the Spiritual World like *Religion The Eternal Dilemma* have been reviewed. In the Occult World, *The Interpretation of Dreams* finds a place.

To encourage the reader to probe further, a bird's eye view of some the interesting aspects of the book is given below:

Did you know that the human body is made up of 7 *octillion* or $7^{27}$ *atoms* the equivalent of 37 *trillion* or $37^{12}$ *cells?* or that the human body *is a kingdom of cells?* That *cells constantly die and get renewed seamlessly providing substance and life and remarkable capabilities to innumerable body parts from birth till death?*

*Your eyes can distinguish between 2.3 and 7.5 million different colours,* and *your nose can identify up to 1 trillion different scents and odours.*

A bewildering fact relates to the seminal cell of the human body, its *DNA* (deoxyribonucleic acid).

*If all the DNA in an individual body is uncoiled it would stretch for one billion miles from Earth to the most distant planet in our solar system, Pluto and back.*

The *Big Five* organs, vital for good health and well being are described. The listing includes the heart, the brain and *mind*, the lungs, the liver and the kidneys.

In respect of glands, the Endocrine Glands get special attention. Most of us have only a cursory knowledge of the Endocrine Glands like the *thyroid* and the *pituitary* that are *ductless* unlike Exocrine Glands like the *mammary* and *salivary* that discharge their secretions into the blood stream through ducts. Glands play a vital role in regulating growth and development of the human body from birth through childhood, adolescence, parenthood till death.

**Part II** relates to the fascinating world of Math Conundrums.

An assortment of topics like *The Power of the Exponent, The Origin of Zero, Euler's Identity, The Ramanujan Magic Square, Chaos Theory, The Seven Bridges of Koningsberg* and others considered as a mathematician's delight have been included.

**Part III** is the concluding part of the book captioned as *Knowledge Frontier Areas.* It covers a wide range of topics that are poised at the outer fringes of our knowledge horizon. They are constantly in the spotlight and subject to frequent review and refinement. Established theories on the *Big Bang, Origin of Life, The Speed of Light, Dark Matter and Dark Energy, The Three Body Problem* and refinements in counter intuitive phenomena like *Quantum Entanglement* find a place.

The chapters in the book constitute a collection of independent essays that cover a wide medley of thoughts. It is *not* a story to be read from cover to cover. The reader is left with the freedom to read the contents in the order of his individual preference.

I do hope that those interested in the *abstract* will enjoy reading my book.

# List of Illustrations

# PART I

## THE HUMAN LIFE FORM

# One

# The Miracle of Conception

*The fertilization of an egg by a sperm cell is one of the greatest wonders of nature. If it were a rare event, or if it occurred in some distant land, our museums and universities would organize expeditions to witness it, and newspapers would record its outcome with enthusiasm.*

**– Dr. George W. Corner, Embryologist at the Rockefeller Institute**

I have referred to conception as a *miracle.* It is nothing short of that.

What is a miracle?

There are many definitions. The one explained below is the most appropriate to our context.

In common parlance a miracle is an occurrence *unexplainable* by existing natural laws. Most of us extend the definition to include a *divine interface* that precludes discussion and debate.

I have avoided this definition.

Instead, the following definition meets the requirements of this essay:

*A miracle is an amazing or wonderful occurrence.* It is derived from the Latin word *miraculum* or *object of wonder.*

Well wishers anxiously waiting outside the labour room of a maternity hospital are ecstatic when this *object of wonder* is successfully delivered. But we are talking of the end result, putting the cart before the horse. Few pause to think about the remarkable processes that *precede* the phenomenon of childbirth.

Modern science enables us to peek into what goes on inside the the body of a woman on the verge of getting impregnated, and the mind boggling occurrences within her reproductive system that lead to it.

For a deeper understanding of the subject I am digressing from the topic of conception in humans to an equally spectacular event that occurs periodically in the marine world i.e.

*The spawning of the salmon fish.*

This comparison gives us an idea of the fantastic range of life forms on earth and their reproductive processes that enable their long term survival.

The *spawning of salmon* is mainly confined to the two biggest oceans on our planet, the Pacific Ocean that connects the western coastline of the American continents with the Far East and Asia. The Atlantic Ocean connects the eastern seaboard of the Americas to the land masses of Europe and Africa.

The Atlantic Ocean has only one species of salmon, the *Salmo salar* in contrast to the Pacific Ocean that breeds several species. I have chosen the salmon to highlight how nature requires other life forms to go through multiple hazards in order to reproduce. It seems to be Nature's way to enable the millions of life forms on planet earth to seek out a *survival pathway* that optimizes their chances for life and continuity through reproduction. The humble salmon, does not qualify for the top billing in marine life forms, either in terms of size or as a food item. In terms of size, the winner is the prodigious *blue whale* weighing 200 tons (equivalent of 33 elephants) and a heart as large as a modern automobile. The blue whale *whistle* exceeds the stinging screech of a Boeing 747 during take off and landing. It can be heard more than 100 miles away.

As a food item, the salmon ranks below the *tuna* arguably the most widely eaten fish in the world. Despite the modest ranking, I have chosen the salmon to contrast with the human life form in respect of reproduction, more specifically *conception* that brings new *lives* into our world. The Atlantic salmon spends its early life in fresh water lakes like Lake Ontario and rivers like the St.Lawrence in the USA before migrating to the brackish waters of the Atlantic Ocean where they grow and mature. As a part of their life cycle they return to their river of origin to *spawn.* The male and female in enormous shoals concentrate their entire energy on reproduction during the late fall/early spring months. The female lays pea size eggs hidden in the loose gravel of the riverbed. The male for its part discharges its seminal fluid or *milt* over the eggs to fertilize them. The success rate is very low (less than 2%) and as a safeguard the female lays several million eggs. The cycle repeats - the parent salmon heads back to the ocean only to return to the river in the fall for the *next spawning season.* The Atlantic salmon repeats this cycle several times in the course of its lifespan. Strangely, the Pacific salmon is able to withstand *only a single spawning cycle.*

We now revert to arguably an equally complicated process within the human female body that precedes conception.

Of the trillions of cells in the human body, we zero in on the *largest,* the female egg or ovum barely visible to the naked eye and the *smallest,* the male sperm that is invisible. The woman's reproductive system depends on two ovaries that produce one egg per month aggregating to 400 -500 eggs in her productive lifespan from puberty to menopause. The liberated egg or ovum that gets released from either of the two ovaries is a tiny shining sphere, two million of which could fit into a tablespoon. The egg has just twenty four hours to get fertilized failing which it will die. During this period it faces enormous challenges to its existence as it drifts from the ovary to the funnel like mouth of the *Fallopian tube,* an important part of the female reproductive anatomy. The muscular movements within the tube propel it to a position where it stands a chance of a rendezvous with the male sperm and get fertilized.

The odds against this happening are so high that the ejected semInal fluid is endowed with as many as 500 million sperm cells. Occasionally, one of these wriggling cells is lucky enough to survive its perilous journey from the vagina to the Fallopian tube and double lucky to latch on to the ovum and fertilize it.

Once fertilized, the egg completes its journey to the womb wall and gets firmly embedded in it. Its rations of starch, protein, and sugar will last only for a week, after which the fertilized egg gets parasitical and will depend on the resources of the mother till delivered as a full grown human child.

The life sketch of the male sperm cell is no less remarkable. It measures 1/500 of an inch and is invisible to the naked eye.

Every single one of the 500 million present in the seminal fluid is a veritable *bundle of energy*. It has a small oval head and a nine times longer *thrashing whiplike tail* that propels it along like the whirring blades of a helicopter. It has to complete a perilous 5 inch journey from the vagina to the farther end of the Fallopian tube to reach the waiting ovum and complete the miraculous act of fertilization. *The sperm's five inch journey to the egg is the equivalent of a man's five mile swim upstream.*

Most of them perish during the journey and a few lucky ones get close to the ovum.

Reverting to the salmon analogy, the inherent perils in the human reproductive process and the salmon instinct to *spawn* is exemplified by the Pacific salmon. The matured salmon shoals in the Bering sea swarm into the mighty Yukon river across Alaska in the USA to the *spawning river beds* in the state of British Columbia in Canada.

The salmon shoals cover a mind blowing distance of 2000 miles (half the width of Canada) in its irresistible urge to spawn.

Herein lies the clue to the *striking similarity* between the human reproductive process and the spawning of the salmon fish.

*The extraordinary energy that propels myriads of sperm cells from the female vagina onwards to the Fallopian tube, each with the singular purpose of seeking out and uniting with the waiting ovum. The same energy propels the salmon to go through the perilous cycles of mass migration from the brackish ocean to the fresh water rivers with the singular purpose to lay its eggs and hatch them, or to spawn.*

*In both cases, new lives are brought into our world. There are more than 8 million known life forms on planet Earth.*

*Each with a survival and reproduction history no less intriguing than that of the human being or the salmon fish.*

*In almost every case, the propagation of the life form is ensured. This process carries on relentlessly over enormous swathes of time without a single break.*

*The occasional exception, like the dinosaur becoming extinct on our planet took a catastrophic event like an asteroid strike to cause it. Otherwise the life and death cycles go on and on uninterrupted for countless millennia.*

This begs the question:

What is the underlying cause behind the preservation and reproduction of the different *land, avian,* and *marine* species on earth?

*God?*

*Nature?*

*Evolution?*

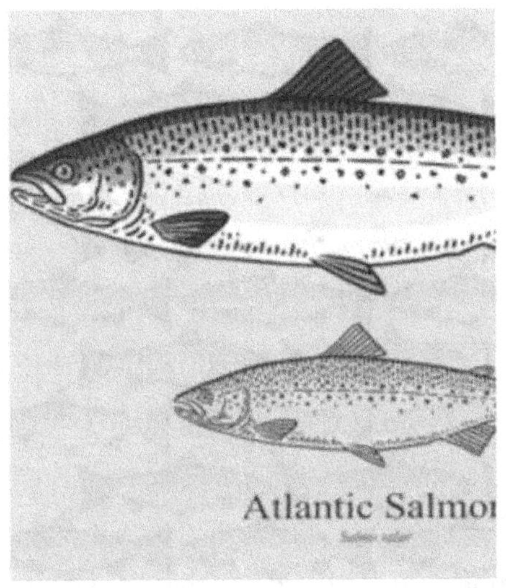


**Illustration 1:** The Atlantic Salmon - Shutterstock _492873088

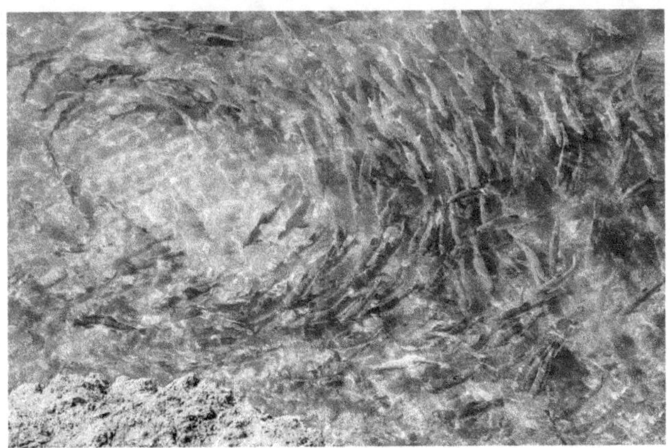

**Illustration 2:** Salmon Shoals in the Pacific- Shutterstock _ 379109818

# Two

# Gestational Period In Humans

## The Mystery of Sex

*The sex of every child is fixed at the instant of conception - from the moment the father's sperm enters the mother's egg, nothing can change what is to be a girl into a boy or vice versa*

**Amram Scheinfeld – Writer on Human Genetics**

First, a few definitions -

**Zygote or pre-embryo:** A fertilized egg that has begun cell division.

In two weeks it becomes an embryo.

In eight weeks the embryo is termed a fetus.

**Gene:** The part within every human that makes him what he looks like on the outside and determines the working inside him. In other words your genes make you different from *everyone alive now* and *everyone who has ever lived.*

It also creates in you traits, habits, and mannerisms that resemble your parents and others in your family tree.

These phenomena arise out of genes being made up of segments of *deoxyribonucleic acid* (DNA) that contain the code for specific proteins that function in one or more types of cells in the body.

**Gamete:** The reproductive or sex cells of any organism, human or otherwise. For humans, the female produces the bigger gamete, the *ovum* and the male the smaller *sperm cell.*

**Chromosomes:** are threadlike structures within cells that contain a person's genes. Each chromosome is made up of a *specific* protein and a *single molecule* of DNA. The largest Chromosome 1, contains about 8000 genes. The smallest, Chromosome 21 contains only 300 genes. Human beings have 23 pairs of chromosomes in every cell.

The female produces only the 'X' type of sex chromosome, the male produces 'X' and 'Y' types in *equal measure*. At the instant of fertilization, depending on who wins the race to unite with the waiting ovum, the sex of *new life* is determined - a boy (XY) or a girl (XX)

The sex factor of the unborn child is a subject that has generated endless controversy and debate. Newlywed parents openly or secretly wish their firstborn to be a male child. This preference, since times immemorial applies across countries, religions, caste, and colour. In India, there is rejoicing on the birth of a male child, disappointment, even mourning if it is a girl child. This preference has persisted for centuries and is beyond discussion and debate - it is a given. Statistics on actual male - female ratios indicate a slight tilt in favour of the male child.

In common parlance we refer to the female as the *weaker sex.* This description is totally misplaced.

In the words of Mahatma Gandhi,

*The female is not the weaker sex. It is the nobler of the two.*

In our present context, from conception onwards, studies have shown the male embryo to be more liable for a miscarriage. Even in respect of still borns and mentally challenged babies, the ratio shows higher number of males.

Yet in the final reckoning, the ratio of normal births indicates 105 males for 100 females. It seems to indicate that the male female ratio at the point of conception is more likely to be 120 to 100 in favour of males to compensate for losses and deficiencies during the arduous journey of the fertilized egg starting at conception through about 9 to 10 months of gestation. This seems odd, since the male sperm cells consist of an equal number of 'X' and 'Y' chromosomes that should on fertilization with the ovum (only 'X' chromosome) yield an equal number of XX (girl) and XY (boy) zygotes.

A possible explanation is that in the race towards the female ovum, the male 'Y' chromosome moves faster than its 'X' cousin and secures a higher number of hits.

We now narrate the fascinating story of the fertilized egg firmly nestled in the mother's womb. The gestational period of nine to ten months for humans is divided into three sections referred to as *Trimesters*.

The First and Second Trimesters, are of 3 months (12 weeks) duration.

The Third Trimester is longer, about 4 months (16 weeks) and the exact date of the baby's arrival varies from one expectant mother to another.

*THE FIRST TRIMESTER*

*THE FIRST, SECOND, AND THIRD*

*MONTHS OF PREGNANCY*

*THE FIRST MONTH*

*Week One to Week Four: A New Life*

It may sound strange, but you are not actually pregnant the first two weeks of the time allotted to your pregnancy. Your gynecologist, by convention counts ahead 40 weeks from the start of your last period, about 2 weeks *ahead* of actual conception. The Chinese reckon the first day of the last period as the *first day of life* of the baby. Further, they consider the tumultuous nine to ten month gestational period as being equivalent to a full year. In the East Asian system of age reckoning, the baby is *one year old* when born. Besides China, this system is still followed in parts of Japan and Korea. Rest of the world follow the Gregorian Calendar coordinates where the baby's life begins *at a defined location on earth and at a precise moment.*

In India the above 2 factors have immense astrological significance -

*During the first stage of labour when the baby's head emerges, it experiences its first impact of radiation from the prevailing planetary configuration.*

It is the first scene of an unfolding drama, the life sketch of a brand new human being that has never been portrayed before.

The first scene identifies the *position of our Natal Sun, the North lunar node* (Rahu), *the South lunar node* (Ketu) *and planets in our solar system.*

*The moment further defines the Zodiacal segment that is rising on the Eastern horizon* (Ascendant)

It depicts the beginning of the lifecycle of a human being who has entered our world. The moments that follow would be indicative of subsequent changing planetary configurations and their impact on the human's life. They will play out as the remaining scenes of the drama. The last scene will depict the termination of the human's life cycle that has progressed through childhood, adolescence, adulthood, parenthood and old age. Hindus refer to the place and moment of birth as determining the *Kundali* or destiny of the baby.

(The above facile astrological summary of a human birth is neither self explanatory nor comprehensive.

Rather, it is an *oversimplification of the impact of the sun, moon, planets and Ascendant alignment* commencing at the moment of a human birth till the end of its life cycle. Astrology is a vast science in itself with an increasing number of adherents across the globe. It is the domain of the professional astrologer in respect of planetary alignments, interpretations and predictions. Be that as it may, the aforesaid brief summary does meet the introductory needs of our present narrative.)

During the first month, the barely visible fertilized egg grows into a human embryo, a quarter of an inch long, increasing *fifty times in size and nearly ten thousand times in weight.*

This remarkable transformation is made possible by cell division or *cleavage.* The cell numbers multiply *geometrically* to form an initial cluster of millions of cells resembling a tiny raspberry. It is called a *morula.* The super active morula, now dividing at a rapid pace is called a *blastocyst.* Within the blastocyst, the inner group of cells is the embryo with a head, a body, and a *tail.* It has a heart that beats and blood that circulates.

The outer layer forms part of the *placenta* that will nourish the baby through pregnancy. Within the placenta, a feeding layer the *trophoblast* forms on the outer edge to pass on nutrients and oxygen from the mother to the embryo.

Beginnings of arms, legs, eyes, ears, stomach, and brain become noticeable.

In the first 30 days, this phenomenon extends to all body parts of a full grown baby. The mother's blood is the source for food, water and oxygen that is absorbed by the trophoblast before passing through blood vessels in the *umbilical cord* that connects the embryo to the mother for the entire gestational period. At no stage does the mother's blood *actually circulate through the embryo.* The embryo itself has gradually changed into a small hollow organ with two cavities separated by a plate called the *embryonic disc that develops eventually into the full grown baby.*

The upper cavity forms the *amniotic water sac* in which the embryo *floats* for the remaining gestational period.

The lower cavity empty sac disconnects from the embryo. Having made sure of its safety, the *double layered embryonic plate can now wholeheartedly grow. The two most important organs, the heart and the brain are the first to develop.*

The rudimentary heart that will undergo many changes before reaching its final form begins to *pulsate at once.* First a twitch, then another, and soon it is rhythmically contracting and expanding, forcing blood to circulate through the blood vessels of the embryo. This movement, once started, continues till birth and thereafter *without a break till the end of the individual's life.*

In a concurrent development, the rudimentary brain and spinal cord begin to form at the two ends of the *neural plate.* By the fourth week, the beginning of the human nervous system that will give man mastery over all other life forms begins to take shape. Next, it is the turn of the stomach or gut to develop in the embryo. The flat embryonic disc becomes humped up in the middle into a long ridge like pocket with openings at either end. The foregut opening forms the mouth whereas the hind gut opening remains closed for sometime. By the twenty fifth day, the embryo can be described as a *small creature about one tenth of an inch long with head and tail ends, a back and a belly. The arms, legs, face and neck are mere stubs that have begun to grow. It has a heart close to its brain.* Within this rather ugly exterior, he has started to form his *lungs* that first appear as a shallow groove in the floor of the foregut. His *liver* begins to appear as a thickening in the wall of the foregut, behind the heart. Heading in the direction of the hind gut there is a long devious path that will ultimately lead to the formation of his *kidneys.* By the end of the month, the embryo is about one fourth of an inch long, curled almost in a circle, with a short pointed tail below his belly. In the head the *eyes* have arisen as two small pouches thrust out from the young brain. The skin over the front of the head shows two sunken pouches of thickened tissue, the beginning of a *nose*. At a short distance behind each eye an *ear* has started to develop.

In thirty days, the simple ovum and sperm cell have mysteriously combined and evolved into a life form destined to become the forerunner of humanity.

## THE SECOND MONTH

### Week Five to Week Eight:

### Development of a Face

From tadpole to man - so one might characterize the changes that take place during the second month of life.

In this month the embryo increases its size to *one and a half inches*. Its weight increases *five hundred fold*.

The eyes that were on the sides of the head, shift to the front, the eyelids develop but remain closed. The limb buds elongate, the ends flatten to a paddle like ridge with five shallow grooves to form the finger plate and the toe plate. Constrictions within each limb mark off the *elbow, wrist, knee,* and *ankle.* Strangely, the human tail reaches its greatest development in the fifth week after which it gradually regresses.

In respect of bone development, a pattern of the bone is first formed in cartilage. As a sculptor first fashions his work in clay, and when satisfied with the design, casts the statue in bronze, so the developing embryo seems to plan out its skeleton in cartilage and then cast it in bone. This process continues through every month of life before birth, through childhood, adolescence till full maturity.

Development of the sex organs is intriguing. By the end of the month, the sex is clearly evident in the internal sex organs and begins to manifest externally.

*The second month closes with the stamp of human likeness clearly imprinted on the embryo.*

In the next seven months the young human being is called a *fetus* that thereafter undergoes growth and detailed development.

*THE THIRD MONTH*

*Week Nine to Week Twelve:*

*The fetus begins to move.*

The male embryo asserts its superiority over the timid female for the first time. It actively plunges into sexual development unlike the female that seems to prefer sexual indifference.

It can be labelled as the tooth month, since sockets form in the hardening jawbone from which the buds for temporary teeth begin to emerge.

Incomplete *vocal cords* that six months down the road will cause the first cry of the baby are beginning to form.

The fetus floats in the cushion of amniotic fluid. It gets its first lesson in *swallowing* by the ingress of a minute quantity of fluid, and expulsion of unwanted fluid from its lungs.

This practice will stand it in good stead to absorb oxygen and food from the mother's blood *right up to the point of child birth*. The digestive system of the fetus now gets underway. The cells lining the stomach begin to secrete mucus to lubricate the passage of food.

The liver begins to pour bile into the rudimentary intestines.

The kidneys begin to secrete urine that seeps out of the fetal bladder into the amniotic fluid.

In the face, the developing jawbones, cheekbones, and the nasal bone forming the bridge of the nose begin to form. Human contours form for the first time on the *small wizened fetal face*.

The developed muscles produce spontaneous movements of the arms, legs and shoulders, and even of the fingers.

*THE SECOND TRIMESTER*

*THE FOURTH, FIFTH, AND SIXTH*

*MONTHS OF PREGNANCY*

*THE FOURTH MONTH*

*Week Thirteen to Week Sixteen:*

*The Quickening*

Death casts its shadow over man *before* he is born, for the stream of life flows most swiftly through the embryo and young fetus and then *slows down* within the uterus. The period of greatest growth occurs during the third and fourth fetal months when the fetus grows *six to eight inches in length*, about half its length at birth. Thereafter the rate of growth decreases steadily.

At two months, the gnome like creature has a head almost one fourth of the body; from the third to fifth it is one third, at birth one fourth, and in the grown adult one tenth of the body height. At four months, the fetus is reasonably good looking with a resemblance to a normal infant.

More work needs to be done.

At the tip of each finger and toe, patterned whorls of skin ridges appear, the basis of future *fingerprints* and *toe prints*. Each human being is marked for life with an *individual, unchangeable stamp of identity.*

From the fourth month, the fetus stirs, stretches, and vigorously thrusts out arms and legs, the *quickening in the womb* as perceived by our ancestors.

## THE FIFTH MONTH

*Week Seventeen to Week Twenty:*

*The straightening of the body axis*

In the fifth month of the gestational period, the surface of the skin of the fetus gets covered with tough dead cells that forms a protective layer for the soft tissues that have formed beneath. *Sebaceous glands* for the first time secrete oil at the base of each hair follicle. This forms a cheesy paste with the dead skin cells that gradually covers the entire body of the fetus. This is in addition to the sloughing off of old skin cells by new ones as in a living being laying the foundation for the *skin, nails*, and *milk teeth*. Important as these developments are, they are overshadowed by the remarkable change in the shape of the fetus that occurs in the fifth month.

In the second month, the embryo forms almost a *closed circle* with its tail not far from its head. At three months the head has been raised considerably and the back forms a *shallow curve*. At five months, the *head is erectly balanced on the newly formed neck and the curvature of the back reduces further.*

At birth the head is perfectly erect and the back is *unbelievably straight.*

In fact it is more straight than it will ever be again, for as soon as the child learns to sit and walk, *curvatures appear in the spinal column* that enable body balance.

How would one describe the five month old fetus?

A one foot long lean creature with wrinkled skin weighing about half a kilogram. If born it may live for a few minutes, take a few breaths and may cry. Although able to move its arms and legs actively, it is as yet unable to maintain the complex movements necessary for continuous *breathing* and will die.

## THE SIXTH MONTH

*Week Twenty One to Week Twenty Four:*

*Eyes that open in the darkness.*

During the sixth month the eyelids, fused shut since the third month reopen.

Completely formed eyes are now visible that will respond to light in a month's time.

Within the mouth, *taste buds* emerge on the surface of the tongue, and on the roof and walls of the mouth and throat.

The six month fetus, if born will breathe, cry, squirm, and may live for several hours. Chances of long term survival are extremely slight unless it is protected in an incubator. The vitality to live is very weak and can be easily snuffed out with the first contact with the external world.

## THE THIRD TRIMESTER
## THE SEVENTH, EIGHTH, NINTH, AND
## TENTH MONTHS OF PREGNANCY
## THE SEVENTH MONTH

*Week Twenty Five to Week Twenty Eight*

*The dormant brain.*

The fetus gets its first taste of independence. Although it needs to spend two to three more months in the uterus, It seems to know instinctively that there is a fair chance of survival if now exposed to the external world.

The nervous system is the last to develop.

This is crucial since parts of the brain that regulate *constant rhythmic breathing, muscular contractions when swallowing,* and the *intricate mechanisms that maintain body temperature* need to function seamlessly for long term well being. By the third month of life, special regions and structures have developed within the brain - the *cerebellum* that receives fibres mostly from the ear. There are also two large sacs, the *cerebral hemispheres.* They are destined to become the *most complex and elaborately designed structures in any life form.* It will provide man unquestionable dominance over all other life forms on our planet.

The remarkable seven month old fetus is a red skinned, wrinkled, child with the look of a *wise old man.* It is about sixteen inches long and weighs about three pounds. He can perceive the difference between light and darkness and has a good chance to survive if

born.

*Eighth, Ninth, and Tenth months*

*Week Twenty Nine to Week Forty*

*Beauty that is skin - deep.*

This period could well be described as the cosmetic phase. Fat is formed rapidly all over the body of the advanced fetus smoothing out the wrinkled flabby skin and rounding out the contours. The dull red colour is replaced gradually by a flesh pink shade. The fetus acquires the lineaments of a human infant.

Pigmentation of the skin is marginal. Even the offspring of coloured races are light skinned at birth. This is particularly noticeable in the iris of the eyes. Regardless of race, the eyes of most infants are *blue - gray* at birth.

The activity of the fetus within the womb now increases several fold. He thrashes out with arms and legs and even changes his position in the rather restricted area he finds himself in.

The proverbial *nine months and ten days* to define the gestational period is not sacrosanct. However as per available statistics 10 percent of fetuses are born on the 280th day after the onset of the last menstrual period and 75 percent are born within two weeks of that day.

The newborn baby usually weighs between 3 and 3.5 kilograms and has a length of 19 inches. Its arms are folded across the chest and thighs drawn up against the stomach to optimize the limited space. This is a small but complete human being. He is on the threshold of facing his first great ordeal, *the process of being born.* The climactic great event is about to happen. Yet all would agree that his experience in the preceding nine odd months has been *truly marvelous.*

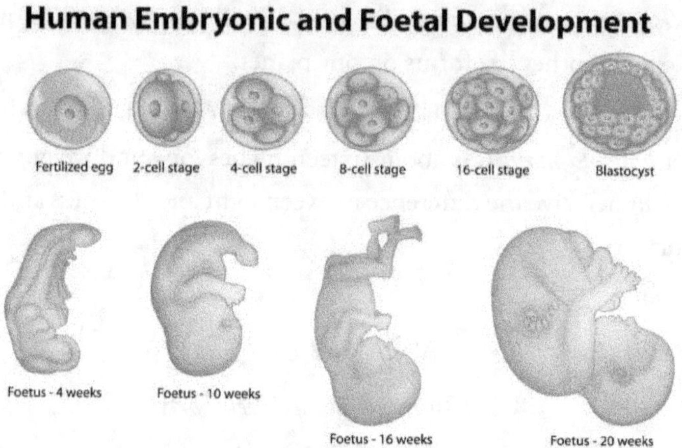


**Illustration 3:** Human Embryonic and Foetal Development - Shutterstock_139691557

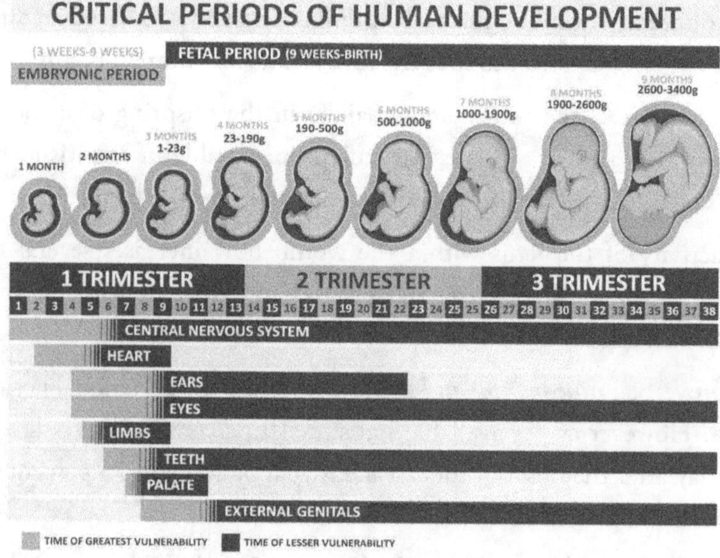


**Illustration 4:** Critical Periods of Human Development - Shutterstock_1616044135

# Three

# Nature's Greatest Event Childbirth

*The pain of childbirth is not remembered. Its the child that is remembered.*

**Freeman John Dyson (1923 – 2020)**
**British American Theoretical Physicist**

At each tick of the second hand of your watch a baby is born somewhere in the world - an event *so frequent that much of its wonder is lost.* The scientific mind wonders. The investigator is forced to admit that there are great gaps in our understanding of this commonplace event. *Indeed, we do not even know why a baby is born when it is.*

Why does the mother suddenly decide to rid itself of an infant she has sheltered for nine long months?

What causes the dramatic changes that take place in the bodies of the mother and child at the moment of birth?

How does the newly born infant take its first gasping breath on which its life depends?

We do not have all the answers. However, a great deal of knowledge based on first hand experience from maternity hospitals across the world has accumulated. It starts with a small pain caused by contractions of the uterus, entirely involuntary but an unmistakable signal to the mother what lies ahead.

(In *Pearl S Buck's* celebrated book *The Good Earth*, the village woman in China, *Olan* leaves the field where she is working, goes home, delivers her baby and returns to the field to continue her work in an uninterrupted smooth flow)

The woman's body has prepared for the great event in many ways -

Her heart has enlarged slightly and her blood volume is up by 25 percent to provide for the baby to come. Her breasts have enlarged to provide for an intricate network of milk ducts and new blood vessels. The mouth of the uterus hitherto hard and fibrous becomes soft and pliable. The *cervix* the opening in the uterus is normally the size of a soda straw. It will have to enlarge to five or six inches to permit passage of the baby.

The birth canal tissues have begun to secrete a starchy stuff called *glycogen* that changes first to glucose and next to *lactic acid*. The acidic environment assists in warding off bacterial infection.

Proper hygiene during childbirth is very important.

(The *Cry and the Covenant*, by Morton Thompson describes the widespread infant deaths in Europe in 1930 caused by poor neonatal hygiene leading to bacterial infection termed *puerperal fever*)

The hormone *relaxin* that is released prior to childbirth relaxes ligaments and tissues in the pelvic area. At the time of birth, the stretched muscle fibres in the upper portion of the uterus begin to contract. This movement pulls the lower uterus, normally passive over the baby's head. Ninety six percent of babies are born in a *head down* position. Only about 3 percent fall into the *buttocks down* alignment. The treating obstetrician corrects the abnormality in the eighth month of pregnancy.

The first contractions, push the head of the baby against the cervix. Each pressure dilates this tiny opening a little more. Dilating it to its greatest diameter is a trying and time consuming process.

The first baby on an average takes sixteen hours, eight to ten hours for subsequent deliveries. Generally, the amniotic sac, the *bag of waters* enclosing the baby breaks when the cervix is fully dilated.

The second stage of labour now begins. The baby passes through the cervix and finally the *perineum*, the muscle bed at the bottom of the pelvis. To facilitate the passage, the baby's head bones are *not* knit firmly together. This explains the distressingly misshapen heads of many newly borns. Fortunately they regain their normal shape within a week.

The average time to complete the second stage of labour varies from forty five minutes to two hours depending on the mother's history - the first born taking the longest time.

The most critical point in human life is the *moment of birth*. The lungs thus far have been useless - tiny unused collapsed balloons since the baby obtained his oxygen ration from the mother. Once born, the baby has to fend for itself. The first breath triggers an intricate series of events -

Since the baby's lungs gathered no oxygen in the watery uterus, there was no blood supply to its lungs. Blood got diverted from the right side to the left side of the heart instead of being sent to the lungs to pick up oxygen. At the point of birth, this comfort level of the unborn infant is suddenly snatched away.

The tiny opening in the baby's heart closes and blood is sent to the lungs for the first time. If the hole does not close we have the phenomenon of a *blue baby* being born. The practice of cutting the umbilical cord as soon as the baby is born has been discontinued. It is now realized that there is a final surge of blood from the placenta into the baby's body. The modern obstetrician waits for this last ration of blood before cutting the link with the mother.

After delivery of the baby, the mother enters the third stage of labour usually lasting for half an hour. The uterus after a brief rest begins to contract again to expel the placenta and simultaneously close off exposed blood vessels.

*This then is the process by which a new human life emerges into our world.*

*Despite its universality, it is without doubt the supreme miracle.*

Globally, about two and a half million infants die either at childbirth or within 28 days of being born. India recorded 522 (in thousands) deaths in 2019. Strangely, the numbers are low in China (64) and Africa - Ethiopia (99), and Tanzania (43)

At birth, something, no one knows what, prompts the baby to take his first vital breath. From then on, the respiratory centre in the brain must *trigger each succeeding breath*. Usually everything goes on without hitch. However, in a distressing number of cases babies do not breathe properly and become victims of *neonatal asphyxia*.

There are many reasons for newborn deaths amongst which indiscriminate use of obstetrical analgesia and other pain killing drugs like barbiturates figure prominently. The medical fraternity recognizes this as a major threat facing newborn babies and is addressing it.

Possible damage to the baby does not end with breathing difficulties. Ailments such as *epilepsy, cerebral palsy and other convulsive states* could be traced to oxygen hunger at birth. Scientists have conducted extensive testing on rats and guinea pigs that confirm the above suspicions.

Gynecologists and obstetricians are coming around to a middle path. They believe in telling prospective first time mothers that there is no such thing as a completely safe method of negating childbirth pain. At the same time a liberal use of analgesia and anesthesia may endanger the life or health of the baby. They encourage them to face as much pain as possible without seeking sedation. With this guideline many women will face the initial stages of labour with minimum sedation.

The use of *Infant Resuscitators* has proved very effective. These range from tiny face masks that administer oxygen to elaborate heated bassinets with intricate oxygen apparatus.

We can and should do more to reduce the hazards of the first day of life.

Women need to train their minds to consider childbirth not as undergoing pain but *doing work.*

With an all round effort by hospitals, gynecologists, obstetricians and expectant mothers a greater proportion of pregnancies will have a happy ending and ensure that *Love's Labour is not lost.*

# Four

# The Baby

The small squirming object is a newborn baby. He merits attention for *there never has been* and *never will be another baby exactly like him.* He is unique in every sense of the word. On an average he weighs 3 kilograms and is 19 inches long. He looks top heavy and is so. His seven pounds is almost entirely in his disproportionately large head and abdomen. There is an open spot in his skull called the *fontanel* covered by a tough membrane that protects the brain lying below it. In many ways the baby is helpless. He has however lived through a great deal in the preceding nine to ten months and developed an inner toughness. Except for crying, yawning, and sneezing being performed for the first time, he has been practicing his entire repertoire of movements for months which his mother will testify to.

The newborn baby has to cry within a minute or two after birth in order to *start breathing air.* The cry is an emergency gasp, a bellows like action of his diaphragm that sucks air into his lungs and drives the fluids out of his nose and throat. The ensuing sound marks another important event - his vocal cords are used for the first time.

His tiny brain soon realizes that crying brings help, and so he will resort to it in the form of shrieks, whines, and grunts intelligible perhaps only to the mother.

Besides, he is now able to grimace, smile, and scowl that may or may not portray emotions normally associated with such facial forms. He hates to have his hands held against his sides. He will spontaneously struggle to free himself from this position. His grasping ability for one so young is extraordinary. He will grasp a rod so firmly that he would rather hang on than let go when lifted. He can *blink* but does so rarely. Perceiving light is his greatest ability at this time, although recognizing familiar objects will take another two months. Of his five senses, taste is the best developed. This explains a baby's tendency to put any object instinctively into its mouth. This finished product that started as a single cell with inherited genes have made him what he is when he enters the world as a brand new human being. Whether this is an *accident* or *part of a design* we will never know. Depending on the genes he inherited, he could well have been born as a dog or cat or any other living being. But he is a human that has entered our world to fight the battle of life that lies ahead. How does a human differ from other life forms? The newborn baby we have been referring to has an unparalleled capacity for development. In the first year his rate of learning is slightly inferior to that of a baby chimpanzee. From then on the contest is over. After the first year he will *race ahead to a privileged group where no other creature can follow.* His power to perceive and act will grow for decades, and his power to understand will keep increasing till the day he dies. At his peak, his brain will be able *not only to assimilate an infinite variety of ideas but to arrange them in patterns and draw conclusions.* He will have unquestionable power over all other life forms and control the destiny of our planet.

*The newborn baby is an extraordinary living object. Nobody knows how the process from embryo to a fully grown baby unfolds in a mother's womb in such large numbers all over our planet.*

*God?*

*Evolution?*

*Maybe either.*

*Maybe neither.*

# MAJOR ORGANS AND GLANDS OF THE BODY

## Five

## The Human Lungs

### WONDERFUL WINDBAGS

*Each organ is related to an emotion, and the lungs are related to grief. When you clear your lungs, you eliminate grief and sadness*

**Amanda Ingber - American author and Yoga Instructor**

Who amongst us has not at some stage in his/her life breathed a sigh of relief?

Our pair of lungs enable deep inhalation and exhalation of air in a crisis providing great relief. Deliberate deep breathing in a calm atmosphere is the essence of the yogic practice of *pranayama* literally meaning control (*yama*) of the life energy (*prana*). Apart from providing relief in a crisis, our pair of lungs qualify as one of the biggest organs in the body located strategically in the right and left of the chest cavity. They are comprised of microscopic sacs called *alveoli* where the phenomenon of *gas exchange* takes place every moment of our lives. Through millions of tiny blood capillaries, inhaled oxygen is exchanged for exhaled carbon dioxide i.e. *blood purification or oxygenation* vital for life sustenance is achieved and maintained. The complexity of lung structure is exemplified by the alveoli. If these minute sacs are placed alongside and spread out, the surface area would cover *an entire tennis court.* Instead of visualizing our lungs as two big balloons, consider them as two buckets of blood with air bubbles within constantly circulating. It is an amazing fact that our lungs contain *as much blood as the entire rest of our body.* We inflate them *half a billion times* in a lifetime of three score and ten. This magnitude of wear and tear would destroy any man made gadget within a few days or weeks. In contrast our lungs provide most of us a lifetime of trouble free service. Cut through, they resemble the cross section of a rubber bath sponge. They are the most important organs in the *Human Respiratory System* and are located at its final end point. For a comprehensive understanding of respiration we need to review the respiratory organs that are positioned at a higher level and provide a pathway to the lungs. Each lung has a duct that connects with the windpipe; it enters near the top and branches out into the organ like a tree. These branches are the *bronchial tubes.* Their job is to deliver inhaled air through 750 million microscopic air sacs called alveoli to the functioning part of the lung. Each alveolus is covered by a cobweb of capillaries, so tiny that red blood cells need to pass through them single file. Through these gossamer walls, blood gives up waste carbon dioxide and absorbs refreshing oxygen. This life sustaining respiratory process is called *gas exchange.* In addition, ancillary processes like optimizing the moisture and temperature of the inhaled air, and protection from the entry of harmful substances take place. Such natural processes are enabled through familiar body responses like coughing and sneezing. Respiratory organs located *above* the lungs are:

*Nose* - a uniquely designed organ, the starting point for respiration. Made mostly of pliable cartilage, it can be mashed and pummeled and still continue to function. Hairs in the lining of the nose prevent entry of dust particles.

*Sinuses* - Hollow spaces in the bones above and below the eyes connected to the nose through small openings.

Sinuses help regulate the temperature and humidity of the inhaled air.

Warming is achieved in the deeper nasal passages where the bones are covered by tissues with rich blood supply. Air passing over these tissues is warmed like air passing over a radiator. On cold days the blood vessels dilate to produce more heat. On warm days they shrink. The nasal air passages have a remarkable sweeping system. Microscopic *cilia* or thin hairs cover the entire route, flailing back and forth, twelve times a second and moving debris in one direction only, towards the throat. Swallowed, the debris is rendered harmless in the acidic digestive tract. The incredible energy of the cilia can be demonstrated by snipping a piece of tissue from a frog's throat. If placed on a table the cilia will *walk* the tissue off the table. If placed inside a bottle, the tissue will *climb* out.

*Mouth* - Apart from the nose, air also enters through the mouth for those with a mouth breathing habit or where nasal passages are temporarily blocked by a cold, or during heavy exercise.

*Throat* - It collects incoming air from the nose and mouth and passes it down to the windpipe.

*Windpipe or trachea* - The passage leading from the throat to the lungs. It divides into two bronchial tubes one for each lung. The bronchial tubes split further into minuscule tubules called *bronchioles*.

*Lung Lobes* - Our right lung is divided into three lobes or sections. Each lobe is like a balloon filled with sponge like tissue. Air moves in and out through one opening - a branch of the bronchial tube. Our left lung is divided into two lobes.

*Pleura* - are two membranes, one continuous and the other folded on itself, that surround each lobe of the lung and separate the lungs from the chest wall.

*Diaphragm* - The strong wall of muscle that separates the chest and abdominal cavities. By moving downwards, it provides suction in the chest drawing in air and expanding the lungs.

*Ribs* - Bones that support and protect the chest cavity. They move slightly to help lungs expand and contract.

*Pulmonary Circulation* - The system in the human body that enables gas exchange. Blood passes through capillaries, the hairlike blood vessels in the walls of the alveoli, entering through the *pulmonary artery* and leaving via the *pulmonary vein*. While in the capillaries, blood gives off carbon dioxide through the capillary wall into the alveoli and takes up oxygen from air in the alveoli.

Every few minutes the body's entire supply of blood must pass through these minute blood vessels. At the start of the pulmonary circulation, dark blue black impure blood flows *in* and bright cherry red oxygenated purified blood flows *out*.

Day and night this life giving process proceeds without interruption. For every heartbeat, the heart pumps an equal amount of blood to your lungs as it does everywhere else in the body.

As mentioned earlier, this process goes on relentlessly without a break for an entire lifetime. It forms the core of the Respiratory System vital for good health and sustenance of human life.

Strangely, the *mucus* that clogs your chest or nose during a cold is the life sustaining lubricant for your lungs. Mucus is a powerful infection fighting agent. Whereas blood actually supports the growth of bacteria, mucus *stops* its growth. Further, if mucus were *not* present in your lungs, you would dehydrate so rapidly that death would follow in a matter of minutes.

At the same time, excess production of mucus can be harmful. Body chemistry provides the right balance.

Whatever you inhale travels from the lungs to the brain in *under seven seconds* e.g. smoke or medicine in vapour form impacts the brain almost immediately. Coughing is not a disease but a protective measure that wards off a sticky piece of food, an allergen, excessive mucus or just food going down the wrong way. Exercise can make your lungs work better and help ward off ailments like asthma. Sadly, vital as they are to human well being, the lungs are vulnerable to a wide variety of diseases including the much dreaded cancer, that is dealt with later in the narrative.

Tips to keep our lungs in good working condition are given below:

**Do Not Smoke*

Cigarette smoking is the main cause of lung cancer and chronic obstructive pulmonary disease (COPD) that includes acute *bronchitis* (infection of the main airways of the lungs, the bronchi that causes them to become inflamed) and *emphysema* (destruction of alveoli).

Cigarette smoke can narrow the air passages and make breathing more difficult. It causes acute inflammation in the lung, that can lead to chronic bronchitis. Over time cigarette smoke destroys lung tissue and may trigger changes that grow into cancer. If you smoke, it is never too late to benefit from quitting. Second hand smoke, chemicals in the home and workplace, and *radon* (radio active gas in the atmosphere) can cause or worsen lung disease. Make your home and car smoke free. Test your home for radon. Avoid exercising outdoors on bad air days.

**The Common Cold:*

A cold or other respiratory infection can sometimes become very serious. There are several things we can do to protect ourselves

# Washing hands often with soap and water or alcohol based cleaners.

# Avoiding crowds during the cold and flu season.

# Good oral hygiene to protect ourselves from the germs in the mouth that could lead to infections.

# Brushing teeth at least twice daily and visiting dentist every six months.

# Getting vaccinated every year against influenza. Practicing social distancing whenever you are sick.

# Being physically active can keep your lungs healthy, whether you are young or old, slim or fat, able bodied or disabled.

Until the 1930s the chest was generally taboo territory for the surgeon, since once it was opened, the lungs no longer in a partial vacuum would collapse and breathing would cease. However subsequent improvements in *anesthesiology -* chiefly the increased use of tubes that can be slipped down the windpipe so that the anesthetist can rhythmically force air and oxygen into the lungs. With this innovation a brilliant new day dawned for chest surgery.

We may spend our entire life in a clean environment, abstain totally from smoking and still get lung cancer.

In such cases, exposure to radiation and genetic factors could be the cause. This is a topic of ongoing research. Of the treatment options other than surgery, chemotherapy ranks high but it's toxic side effects pose serious problems. *Immuno therapies* with decreased side effects are being increasingly used.

Mankind's quest for a safe and fool proof treatment for cancer of the lungs or any other part of the human body continues. Short of total removal of the diseased organ from the body, there is no sure shot treatment.

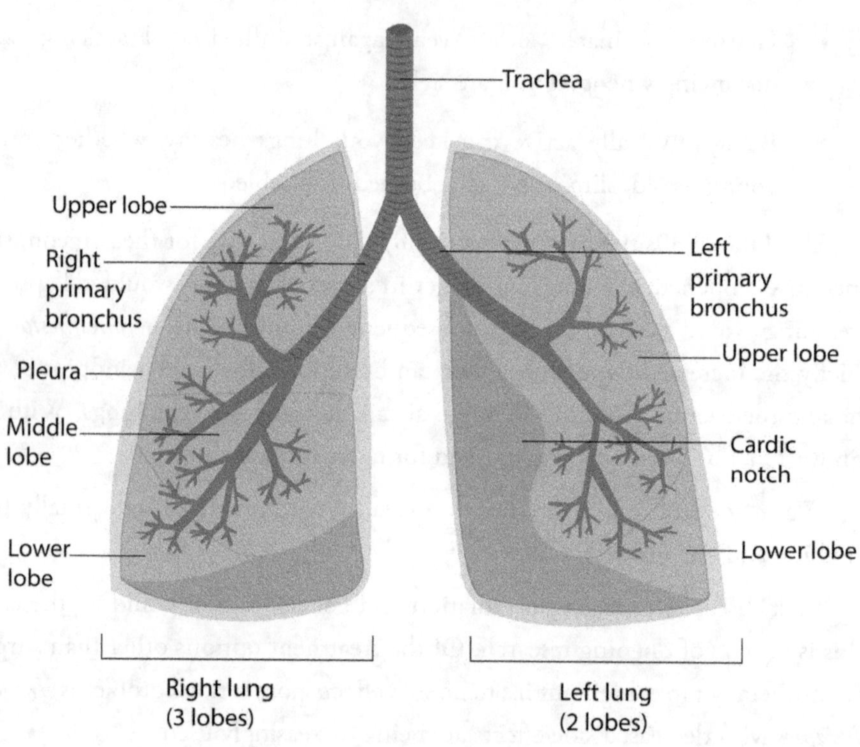


**Illustration 5:** Human Lungs Anatomy- Shutterstock _1823174198

# Six

# Your Liver is Your Life

*I work very hard and I play very hard. I am grateful for life. And I live it - I believe life loves the liver of it. I live it.*

**Maya Angelou (1928 – 2014) American poet and Civil Rights Activist**

## LIVER AS AN ORGAN:

The liver is a large organ, second only to the outer integument of the human body *the skin,* that is heavier and larger.

The human liver is shaped like an angular inverted cone (Sherlock Holmes cap) of dark reddish brown hue, rubbery texture, and weighing about 1.5 kgs. It is located in the upper right hand side of the abdominal cavity to the left of the stomach and below the lungs in the chest cavity. (As mentioned earlier, the diaphragm, a strong muscular wall separates the abdominal cavity organs from the chest cavity that houses the heart and lungs.)

The liver is arguably the most versatile organ in the body - *the only organ that can regenerate itself completely.*

The regeneration happens as long as a minimum of 25% of liver tissue remains. It is amazing that the liver can regrow to its original size and capability *without* any loss of function during the regrowth process. In humans, this process takes only 8 to 15 days which is incredible given the size and complexity of the organ.

## LIVER AS A GLAND:

Besides being a major organ, the liver is also the largest *gland system* in the body. The term *gland* is associated with *exocrine* glands (connected through ducts) to blood vessels like the *sweat, salivary,* and *mammary* glands. It could also refer to *endocrine* ductless glands like the *thyroid* and *pituitary* glands.

In this essay we will attempt to detail the functions of the liver as a major organ and as a major gland both exocrine and endocrine.

What are the differences between organs and glands?

*Glands secrete substances. Not all organs do so e.g. in the abdomen, the spleen, adjacent to the liver has no secretion.

*Glands usually have tubelike structures e.g. gastric glands lining the stomach or the mucus secreting glands in the duodenum of the small intestine. Many organs are *not* tubelike e.g. the liver is a solid dense organ, the stomach a hollow one. The liver as an exocrine gland secretes *bile*, a yellowish green alkaline fluid that contains bile salts. Bile travels from the liver through ducts to the gall bladder for storage. During storage it gets concentrated as much as five times of its original strength. During a meal, the gallbladder releases bile into the small intestine to assist in digestion and absorption of dietary fats. Bile also contains alkaline *bicarbonate ions* that help neutralize acid carried from the stomach. Aside from its exocrine gland function, the liver is also and significantly so an endocrine gland that secretes hormones with diverse biological functions e.g. the liver converts the thyroid hormone into its most active form. This vital hormone is responsible for modulating the body's metabolic rate i.e. the *speed* at which complex biochemical processes and reactions take place. In addition the liver secretes IGF-1 the hormone that promotes cell growth, *angiotensinogen* the hormone that regulates sodium and potassium levels in the kidneys and controls blood pressure.

## LIVER AS A FILTER:

The multifunctional liver with 300 *billion specialized cells* filters about 1.7 litres of blood per minute. The versatile bile helps the small intestine break down food to absorb fats, cholesterol, and some vitamins using complex chemistry.

Anxiety surrounding cholesterol is largely misplaced. It is a waxy, fat like substance that plays a vital role in regulating health. It is found in *all* the cells of the human body. It makes hormones, vitamin D and substances that aid digestion. We obtain our needs of cholesterol through fatty foods in our diet. It is harmful *only* if taken in excess of the body's requirement. Focussing on bile the liver's main secretion, it consists of bile salts, *bilirubin*, electrolytes, and water. Bilirubin is formed by the breakdown of *hemoglobin* in the blood. The iron released from hemoglobin is stored in the liver or bone marrow and used to *make the next generation of blood cells.* Of the vitamins, vitamin K is essential for the production of *coagulants.* If the liver does not produce enough bile, clotting factors in the blood get compromised. Regarding hormonal activity, although sexual urge is regulated in the *gonads,* liver chemistry maintains the right balance of *estrogen* in women and *testosterone* in men to ensure we are neither impotent nor sexually wild. Excess hormones are filtered out of system. The liver also filters out substances like alcohol and drugs. At the cost of repetition, a summary of the wide range of functions carried out by the liver continuously *over an entire lifetime* is given below:

- *Secretion of bile.

- *Assimilation of carbohydrates to maintain normal glucose levels. Excess glucose is stored as *glycogen,* and released whenever a quick burst of energy is needed.

- *Complementing kidney function. (*Amino acids* are metabolized in the kidneys into toxic ammonia. The liver converts ammonia into *urea* that is excreted from the body as urine.)

- *Metabolism of cholesterol and fat.

- *Synthesis and endocrine secretion of growth hormones and coagulants.

- *Detoxification of harmful drugs and poisons.

- *Cleansing of bacteria from blood.

- *Storage of vitamins A, D, E, K and B12. In some cases, several years worth of vitamins is held as backup. The liver also stores *copper* that is released when required by the body.

- *The liver functions as a reservoir for blood.

- *Immunological function: The liver is part of the *mononuclear phagocyte system*. It contains a large number of *Kupffer cells* capable of destroying disease causing agents entering the liver through the gut.

- *Production of *albumin:* Albumin is the most common plasma protein. The liver builds amino acids into the albumin that regulates the salt and water balance in the body crucial for life.

- *Synthesis of angiotensinogen: This hormone raises blood pressure by narrowing the blood vessels when alerted by production of an enzyme called *renin* in the kidneys. One would expect that an organ as complex as the liver and performing such a wide range of vital body functions would be prone to diseases that cause it to occasionally malfunction. Sadly,this is true.

## LIVER PATHOLOGY:

Examples of liver disease include:

*Fasciliasis*: This is caused in tropical regions by the parasitic invasion of the liver by a worm known as a *liver pfluke*.

*Cirrhosis*: Characterized by a process called fibrosis in which normal liver cells are replaced by scar tissue. Fibrosis can be caused by toxins, alcohol, and *hepatitis*.

*Hepatitis*: Acute infection of the liver causing it to get inflamed. Factors like viruses, toxins, or an autoimmune response precede the onset of the infection. In most cases the liver heals itself, but in severe cases liver failure can result.

## Primary Sclerosing Cholangitis:

An inflammatory disease that destroys the bile ducts. The cause is unknown but suspected to be an autoimmune response.

*Fatty Liver Disease*: caused by obesity or alcohol abuse. *Vacuoles* of fat build up in the liver cells. If alcohol abuse is *not* the cause, the condition is called non alcoholic fatty liver disease (NAFLD). If left unchecked it can develop into *non alcoholic steato hepatitis* (NASH) a precursor to liver cirrhosis.

*Gilbert's syndrome*: A genetic disorder that affects a small segment of the population. In this condition, bilirubin is not fully broken down leading to mild *jaundice.*

*Liver Cancer*: Common types are *hepato cellular carcinoma* and *cholangio carcinoma.* The leading causes are alcohol and hepatitis. It is the sixth most common form of cancer and the second most frequent cause of global cancer deaths in 2020 closely trailing lung cancer.

*Prevention is better than cure:*

Factors to bear in mind to keep the liver in good working condition:

*Diet:* As the liver is responsible for digesting fats, excess consumption of fats can overwork the organ. Obesity is linked to fatty liver disease.

*Alcohol abuse:* Compulsive drinking causes cirrhosis of the liver over time.

*Drug abuse*: Millions of teenagers across the world fall prey to illicit non medical drugs. These people can overload the liver with toxins.

## Mixing Medication and Alcohol:

This practice can cause liver failure e.g. taking paracetamol (*Dolo 650*) after a drink.

*Airborne Chemicals Hazard:* When painting or using gardening chemicals, adequate ventilation and use of a mask is necessary. Airborne chemicals can cause liver damage.

## Travel and Vaccinations:

- Protection through vaccination is necessary if you are traveling to an area with known cases of hepatitis A or B.

- Adequate protection from mosquito bites in infested areas. *Malaria* grows and multiplies in the liver. *Yellow fever* can lead to liver failure.

- *Safe sex:* There is no vaccination available for hepatitis C transmitted through sex, tattoos and piercings. **Exposure to blood and germs*:

  Blood transfusion, sharing of toothbrush with friends, and use of dirty needles can cause infection.

## To summarize:

The human liver has remarkable powers of regeneration and can take a lot of abuse with impunity. However this vital organ needs to be protected through judicious lifestyle choices and a healthy diet. Always remember that in respect of your body the liver performs the onerous roles of *policeman and master chemist. This duty, extends for most of us for an entire lifetime without a single shutdown for repair or maintenance.*

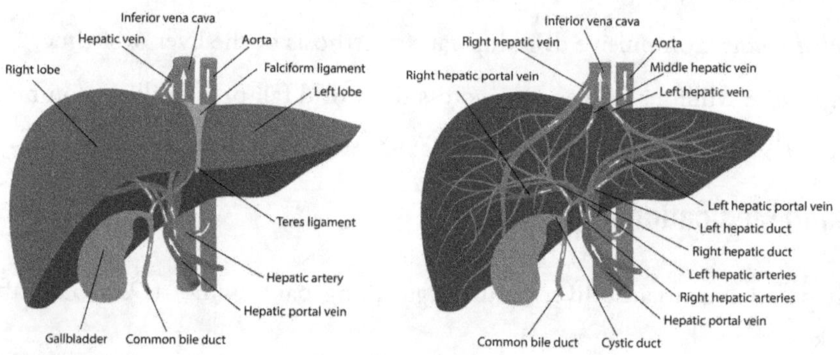


**Illustration 6:** Internal Anatomy of Human Liver - Shutterstock_1315294157

# Seven

## The Human Heart
## Wondrous Courageous Organ

*The human heart feels things the eyes cannot see, and knows what the mind cannot understand.*

**Robert Valett American Professor of Psychology (1927-2008)**

The outline of the human heart drawn on any surface is a symbol of love. If shaded in red it evokes excitement in the viewer. Further, if there is an arrow through the heart the effect gets augmented as an unmistakable expression of love.

The organ itself is a biological marvel; should it stop beating for a few minutes, life would exit from the body. In common parlance, we would say the person is 'dead'!

All primates, including man have a heart and a blood circulatory system. There are a few small living creatures (like jelly fish and flat worms) that do not have a heart. The blood in such creatures does not 'circulate' but 'diffuses' to reach cells at the extremities of the body.

How does the heart work?

Borrow a stethoscope from your doctor and apply it to the left side of your chest.

You will hear the familiar 'lub dub' 'lub dub' 'lub dub'........

This 'contraction' and 'relaxation' that constitute a 'heartbeat' happens 2.5 billion times in 70 years (approx lifespan) without a single shutdown for maintenance!!

The first heart sound (lub) is caused by the contraction and relaxation of blood and the accompanying vibration of the heart at the time of the closure of the "tricuspid" and "mitral" valves. These valves connect the 2 upper chambers of the heart, the "auricles" with their corresponding 2 lower chambers, the "ventricles".

The second heart sound (dub) is caused by the same contraction and relaxation of blood and vibration at the time of closure of the "pulmonic" and "aortic" valves. They complement the pumping actions by the lower chambers, taking deoxygenated blood from 2 major veins of the body, the "superior vena cava" and the "inferior vena cava" to the lungs for oxygenation (pulmonary circulation) and throughout the body (systemic circulation).

Further details on blood circulations by the heart will be provided later in the narrative.

Viewed from the angle that the 'lub dub' sound is crucial in sustaining life, listening to it through a stethoscope can be scary!

The thought arises:

'Suppose it stops' ?

This fear may seem justified, but between 'contractions' the heart does 'relax'; the 'relaxation' time is in fact double the 'contraction' time.!

Also the heart draws more blood as rations for sustenance than warranted by it's size:

1/20 of the total blood in the body, although in terms of weight it is a mere 1/200 of the body!

There is an interesting anecdote on how the first stethoscope was created.

In 1816 Paris physician Rene Laennec was examining a buxom woman with a suspected heart problem. He hesitated to apply his ear on her chest to hear the heartbeat; instead he rolled a piece of paper into a cylindrical form and listened at the other end. He could hear the heartbeat clearly.

The idea of a stethoscope was born!

The heart provides yeoman services to the body as the fountain head of 3 types of blood circulation:

1.  The "systemic circulation" in which bright red oxygenated blood is pumped through arteries to the different parts of the body, and purple impure blood laden with carbon dioxide brought back by veins to the heart in the return journey. To summarize blood "high" in oxygen travels from the heart to different parts of the body, and blood "low" in oxygen makes the return journey to the heart through the veins.

2.  The "pulmonary circulation" surprisingly works in the reverse. Blood "low" in oxygen is pumped by the lower right chamber of the heart through the right and left pulmonary arteries to the 2 lung lobes where it gets oxygenated, and this "high" in oxygen blood travels back to the left lower chamber of the heart through 4 pulmonary veins where the systematic circulation commences.

3.  The "coronary circulation" in which the muscles of the heart are nourished with arterial blood through hair like capillaries. How does this work?

The heart is a tissue like other tissues in the body that requires oxygen and nutrients for its sustenance. Although its chambers are full of blood, they do not nourish the heart itself!Instead, the job is done by "coronary arteries"

Two major coronary arteries branch off from the aorta near the point where it meets with the left ventricle.

The Right Coronary Artery supplies the right auricle and right ventricle. The Left Coronary Artery services the left auricle and left ventricle.

This is a simplistic description. The Coronary Circulation hair like blood capillary network is much more complex. We are omitting these details since they are not required for the purposes of this narrative.

The human heart is indeed a complex, ingeniously crafted organ tasked to work flawlessly for a lifetime without the luxury of taking a break!

Unfortunately a variety of diseases can adversely affect its working capable of causing disablement and death. The simplest would be a massive "cardiac arrest" followed by death of the patient. Such cases are not common. More often than not the defects in the heart do provide warning signals well in advance and are responsive to treatments ranging from medication to surgical procedures. An extreme example of the latter is a "heart transplant" where the patient's diseased heart is replaced by a compatible, healthy, donor heart taken from a terminal patient.

In some cases of CAD (coronary artery disease) where the blood flow to the heart muscle gets restricted due to narrowing of the capillary blood flow, an alternative network normally closed may enlarge and get activated creating a "natural" bypass. This is nature's way of protecting the heart muscle from injury.

When this does not happen, "coronary artery bypass grafting" is resorted to. The cardiac surgeon uses a healthy blood vessel from another part of the patient's body to bypass the blocked artery.

There are many other surgical wonders that regularly happen nowadays in hospitals all over the world:

A surgeon can replace or repair a malfunctioning heart valve.

He can repair congenital heart defects and aneurysms.

Pacemakers, balloon catheters, stents can help regulate the heartbeat and support blood flow.

Trans myocardial laser re vascularization can help treat angina.

The surgeon can create new paths for electrical signals to pass through. This can help treat "atrial fibrillation".

We had referred earlier to an extreme form of surgery called "heart transplant" This requires the use of a HLM (Heart Lung Machine). We shall briefly explain how this remarkable machine works:

The machine is referred to as a "cardiopulmonary" bypass, indicating its function as a means to substitute for the normal functions of the heart (cardio) and lungs (pulmonary).

Surgeons had discovered earlier that they could stop the heart by lowering the patient's body temperature, a condition called "hypothermia" and by flooding the heart with a cold solution. In its state of artificial "hibernation" the body needed less blood circulation, but at best it gave surgeons only a few moments to carry out the surgery.

In 1953, at Jefferson Medical College, Philadelphia, USA a surgeon connected the circulatory system of an 18 year old girl to a new machine, stopped her heart and for 26 minutes operated on her to close a hole in the wall of the heart between the left and right auricles.

It was the first successful use of a HLM and the beginning of a new era in cardiac surgery.

Remarkable as it was at that time, the machine was improved in many ways in the years that followed:

The size of the initial bulky machine was substantially reduced and the quantum of blood required to prime it was limited to a few litres. It became possible to keep the patient alive for a longer time opening up a whole new range of operations. Thanks to the modern refined HLM machine, open heart surgery - especially coronary bypass has become routine throughout the world.

Another major development in the cardiac medical field is the "artificial heart"

This device is quite different from the HLM and in some ways more revolutionary. It has played a major role in rehabilitating heart patients for the past almost 40 years. It has enabled successful heart transplant surgeries in many parts of the world using an artificial heart in cases where a compatible human donor heart option is not immediately feasible.

13 artificial heart designs have been designed, but only one has received commercial approval from the FDA.

During the period of development, designs that were not fully approved served as stop gap relief for afflicted patients in search for a human donor heart.

The Syncardia Total Artificial Heart (TAH) is the sole approved version currently in use in the USA, Canada, and Europe. More than 1800 transplants using this TAH have been carried out.

The successful device, has two ventricles and four valves as in the natural heart. Unlike a donor heart, the TAH is readily available when needed. It is biocompatible with the body and does not require "anti rejection " medications.

The Syncardia TAH is pumped and monitored by an external machine that produces pulses of air and vacuum that pump blood in and out of the ventricle.

The device is made of a highly durable, biocompatible material.

The Syncardia TAH has worked successfully on patients in the age group 9 - 80 years. Patients have lived on the TAH for more than 4.5 years.

The device is available in 50 cc and 70 cc capacities and has been found suitable for men and most women. Stable TAH patients are able to leave the hospital and enjoy active lives while waiting for a donor heart.

Today, the TAH is available at more than 140 hospitals in over 20 countries.

Let us now review some facts relating to the human heart that are so amazing they can scarce be believed:

1. All the blood in your body travels through your heart 'once a minute'

2. Every day, the heart creates enough energy to drive a truck 20 miles. In a lifetime, that is equivalent to driving 'to the moon and back'

3. The heart pumps blood to almost all of the body's 40 trillion cells. Only the cornea in the eye does not receive blood supply.

4. When the body is at rest it takes only '6 seconds' for the blood to go from the heart to the lungs and back, only '8 seconds' for it to go to the brain and back, and only '16 seconds' for it to reach the toes and travel all the way back to the heart.

5. Grab a tennis ball and squeeze it tightly. That's how hard the beating heart works to pump blood.

6. The pressure a human heart generates is strong enough to squirt blood '30 feet' across a room.

A small digression - The heart is arguably the most important organ in the body with the brain a close second.

Is this a case of the 'heart ruling the head'?

The comparison is inappropriate; the heart and brain do not compete with each other- one in fact complements the other! Let me clarify through an analogy:

In a modern airport, the 'head' is the Air Traffic Control, the 'heart' is the aircraft, pilot,crew and passengers. They complement each other and closely coordinate for a safe flight!

What does the heart consist of?

Any objective analyst would agree that it is a 'design masterpiece'

It is often defined as an 'electro muscular double pump' way superior to the most intricate gadget ever devised by man!

It is not difficult to see why this is so. Every minute, the heart pumps about 6 litres of blood to every cell in the human body (approx 40 trillion cells) on a round trip which repeats itself almost 40 million times in the course of an average lifespan of 70 years!

The human heart is a hollow, cone shaped muscle located between the lungs and behind the "sternum" or breastbone. Two thirds of the heart is located to the left of the midline of an individual's body and one third to the right.

The apex points down and to the left. It is 5 inches long, 3.5 inches wide and 2.5 inches from front to back. It is roughly the size of a closed human fist.

The heart has three layers. The smooth inside lining is called the "endocardium". The middle layer of heart muscle is called the "myocardium". It is surrounded by a fluid filled sac called the "pericardium".

The average weight of a female human heart is about 250 gms and the male heart 300 gms which 'hangs' within the rib cage; it is snugly ensconced between the right and left lung and pointing to the left.

It has 4 chambers- 2 upper chambers, the right and left auricles and 2 lower chambers, the right and left ventricles.

The right auricle is connected to the right ventricle by the 'tricuspid valve'. The 'mitral valve' connects the left auricle to the left ventricle.

The right auricle and left auricle can be considered as 'waiting' rooms.

When the valves open, blood flows from the right auricle to the right ventricle and from the left auricle to the left ventricle.

The pumping action takes place in the 2 ventricles i.e. the 2 lower chambers of the heart!

The right side of the heart is separated from the left by a strong muscular wall called the 'septum' i.e. there are no valves between the 2 auricles or the 2 ventricles!

There are 2 more valves in close proximity - they are the 'pulmonic valve' and the 'aortic valve'

We shall describe the functions of these 2 valves shortly!

What goes on within the heart?

The electric part is intriguing!

The AV Node (atrioventricular) is the electrical impulse that signals the auricles to contract.

The SA Node (sinoatrial) in a similar manner signals the ventricles to contract.

The muscular part results in 2 circulation cycles:

1.  The relatively short 'pulmonary circulation' in which the right ventricular contraction pumps the impure blood to the lungs through the 'pulmonic valve'

    The impure blood draws oxygen from the lungs acquired through 'inhalation' and gets 'purified' I

    It simultaneously discharges its freight of carbon dioxide that gets expelled by the lungs through 'exhalation"

2.  The longer 'systemic circulation' where the left ventricular contraction pumps oxygenated blood to the aorta through the 'aortic valve' to different parts of the body!

This is the forward journey where the cells comprising the various organs are rejuvenated.

In the return the blood carries a freight of carbon dioxide and wastes. As explained, the former is expelled through 'exhalation'

The wastes find their way out through the body's excretory organs.

This remarkable activity that commences in the eight week old foetus within the mother's body carries on relentlessly to childbirth, a full life and ceases only at death.

Since the pulmonary and systemic circulations are closed circuits with continuous nonstop flow of blood, there are no start and end points!

Almost all the major blood vessels of the body are connected to the heart.

Examples are the pulmonary artery and aorta to the left ventricle. The great veins i.e. superior and inferior vena cava and pulmonary veins to the right auricle!

The heart repeats the 'lubb dup', 'lubb dup' the 'systole' and the 'diastole' relentlessly in disease and health, through sleep and awaken states, from womb to tomb without ever failing and with an efficiency and courage that surpasses human understanding!!

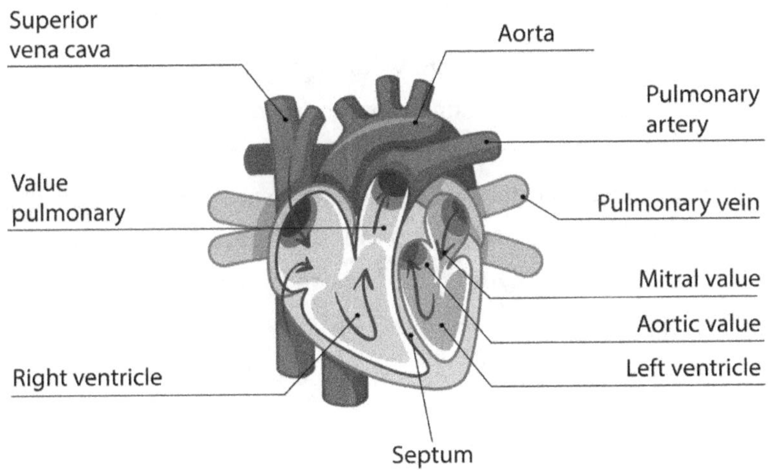


**Illustration 7:** Human Heart Anatomy- Shutterstock_1901690035

# Eight

# The Remarkable Human Brain And Mind

*Biology gives you a brain. Life turns it into a mind.*

**Jeffrey Kent Eugenides American novelist (1960 - )**

Many in the medical profession, would agree that in the human body the big three organs are the brain followed by the heart and lungs. This is not to undermine other organs and innumerable body constituents like glands,blood vessels,muscles,ligaments etc and above all the *trillions of cells* which make up the human body. When all of these are in sync, *the musical melody of the body is in full flow.* There are no discordant notes. The concerned individual enjoys perfect health.

This applies to *all* biological life forms; observed variations are in matters of detail.

Unlike the human brain, which is perceptible to the surgeon during surgery, its close associate, the human mind is a *complete intangible.* It is hidden from sensory perceptions but its presence is undeniable.

I am reminded of the biblical verse:

*The wind bloweth where it listeth and thou hearest the sound thereof but canst not tell whence it cometh, and whither it goeth* (KJV Bible John 3:8)

The mind resembles the wind, invisible but with an undeniable presence!

We shall now focus separately on the brain and the mind, to differentiate *if possible* their distinct roles in shaping the behaviour and action of humans:

1. **The Brain:** Before going into the complex structure of the human brain and its awesome capabilities, let me start with a quote:

*No robot could come even close to duplicating the human brain. A machine even remotely like it would have to be about the size of Rockefeller Centre, and it would take several lifetimes to wire it up. The electrical power requirement would be about equal to the power used now to supply the greater part of New York City. And the necessary cooling system would be so enormous that you would probably have to divert the Hudson to supply it.*

**Dr.Norbert Wiener (1894-1964) American mathematician and philosopher**

The human brain weighs a little less than one and a half kilograms; it is grayish pink in colour,moist and rubbery to the touch with a volume approximately equal to that of a cricket ball. It is enveloped by three membranes and cerebrospinal fluid which absorb bumps and shocks. It sits snugly within the the petrous part of the human skull considered the hardest bone in the body. The different layers of protection of the brain indicate its prime importance as a body organ.

Following aspects of the human brain are truly incredible:

a) The brain never *gets tired!*

When we speak of 'mental fatigue' more often than not the fatigue is somewhere else-eyes,neck, or back. Activity within the brain, unlike most other organs is *electrochemical* and not *muscular.*

b) There are an estimated hundred billion *cells* in the brain-more than ten times the existing population of the world!

Electrochemical messages are crisscrossing these cells continuously; the brain continuously receives messages from sensory organs, and instantaneously issuing commands as responses. Simultaneously other cells are storing new impressions,factual data and sensations in the brain's vast memory storehouse!

c)     Aging does not degrade the brain's *learning ability*. Even if you have missed out on a school and college education, you can still catch up at sixty and beyond. Generally, constant use of the brain as in brainstorming sessions tends to sharpen the brain's ability. Disuse causes it to atrophy.

This phenomenon is to a large extent irrespective of age.

Admittedly, blood circulation does slow down with age, which could deny oxygen and glucose to some cells in the brain causing them to die. This explains why old people can recall events in their childhood but not recent occurrences.

Age *per se* does not cause degradation in brain capability in any significant way.

d)     The *conscious* part of our brain used during day to day living when compared to its overall capability, represents just the tip of the iceberg.

The *subconscious* and *unconscious parts* constituting the balance 90% provides a vast storehouse of information on past events and experiences, a bottomless pit of memory data. Evidence of this manifests itself when a long eluding information unexpectedly reveals itself in a *flash*. Hunches,gut feel and intuition reside in the subconscious.

*Extra sensory perceptions* in some individuals is also believed to emanate from these relatively unexplored regions of the brain!

e)     It is interesting to compare the brain's overall capability with modern super computers. If its a chess game or speed of a calculation the computer would be the winner. Such comparisons do not reveal the bigger picture. In a limited sphere, and in an oversimplified manner, the computer's superiority in such instances is explained by: *Brains are analogue, computers are digital.*

On an overall basis, however the human brain is far more advanced and possesses more computational power than any super computer.

The human brain consists of billions of microscopic neurons which unlike a computer work in *random* rather than *predictable* manner.

A computer, by design has separate parts for processing and memory. The brain does not have watertight compartments; the resultant flexibility provides it with much greater efficiency!

In terms of numbers a hundred neuron transmissions based on biological evolution rules is equivalent to a million steps in a computer with the most modern software. Aside from this, the difference in the *energy expended* by the computer and the brain for performing the same quantum of output is *phenomenal,* something like the power to light an entire high rise building as against a light bulb!

To quantify the difference in efficiency is not easy, because it is not an apple to apple comparison.Researchers in Japan tried to match the processing power in *one second from one percent* of the brain with an advanced supercomputer. It took the 4th fastest supercomputer in the world (the K Computer) *40 minutes* to do a matching job!

f) The brain has three principal structures. The biggest constituting two thirds of the brain is the cerebrum, the centre for reasoning. The cerebellum at the back of the skull comes next which provides balance to the body, and the third is the medulla leading to the spinal column linked to involuntary body tasks like breathing.

Unlike the above three structures, the mind *is not a physical entity.*

2. **The Mind:** The relationship between the mind and the brain goes back to the era of Ancient Greek philosophy-Socrates, Plato and Aristotle circa 500 B.C.E-300 B.C.E

In more recent times French mathematician and philosopher Rene Descartes (1596-1650) identified the brain as the *seat of intelligence* and mind with *consciousness and self-awareness* but could not clarify the dichotomy any further.

In yogic science, the mind was linked not only with day to day consciousness but also the esoteric regions of the *subconscious* and the *unconscious.* Working of the mind was believed to be faster than even the speed of light; the subconscious and unconscious parts of the brain if awakened, were believed to be capable of achieving magical things through the Spark of the Divine.

When probing the nature of the mind there are some surprises:

Body Injury: In the case of serious injury, where a patient lost a limb and suffered excruciating pain in the process, instances were reported where the pain returned with the same intensity later even though the link of the limb with the body had ceased to exist. It seemed to suggest that the mind had a *genetic memory* registering sensations which were no longer registered in the brain. This phenomenon, whose factual accuracy has been established through several case studies,is a subject of further investigation by neurologists.

Impact on physics: Physicists are also contributing in the brain vs mind debate with particular reference to the role played by consciousness in determining the end point of a scientific experiment. The scientist (observer) conducting the experiment involuntarily causes his consciousness to *influence* or even *determine* the nature of the observed at the precise moment of observation. This is a *quantum mind phenomenon!*

Other quantum related examples are wave-particle duality (Max Planck), particle-wave duality (Louis De Broglie), uncertainty (Werner Heisenberg) and superposition collapse (Erwin Schrodinger)

What is going on? To throw light on such weird occurrences in the brain possibly through interaction with the mind scientists have followed a *focused observation approach* I.e.analyze the *core of reality* I.e.break down the microscopic neurons in brain cells into even smaller units and subunits for a better understanding of reality.

Research in this direction reveal that the teeny weeny neuron is itself made up of mind boggling small *micro tubules* (100 micro tubules in a neuron)

Each micro tubule has *virtually countless* number of protein subunits.

It is at the level of these subunits that the secrets of the working of the human mind are possibly hidden!

Doctors who have probed Near Death Experiences are of the view that even after a patient has been declared *brain dead* his mind does not exit but hovers around the deceased for some time!

Mathematical physicists like Roger Penrose and Stuart Hameroff have sought explanations in the nature of existing phenomena like *quantum coherence.*

Earlier, in the eighteenth century, David Hume and Immanuel Kant showed remarkable prescience in providing philosophy based explanations for the laws of classical physics as enunciated by Newton stating that it represented *what was observed* and not what *actually was!*

Spiritualists like Swami Patanjali and M.K.Gandhi have given the topic a spiritual twist.

The closest we have reached to an acceptable understanding of the brain and the abstract entity called the human mind is:

*The brain vs.mind debate may not be a question of either/or after all, but a question of quantum reality, but an interweaving of mind and matter into one which is a simple definition of **yoga**.*

The formidable power and mystery of the human mind has been explained.

The question arises why the brain is given the privileged status of being associated with it and not any other vital organ like the heart or the lungs?

If you study the functioning of the human body, the brain is the *command and control* centre which controls *all* organs e.g. if a dust particle enters your right eye, the sensory receptors in the affected area pick up the signal and pass on to sensory neurons which in turn transfer impulses of discomfort to the *central and peripheral nervous system.* (200 billion neurons constitute the nerve network of an average person) The response comes as a command through motor neurons to the tear glands in the right eye to secrete tears and wash away the offending dust particle (what actually happens involves many more steps within the nervous system but we have provided an abbreviated process that meets the requirements of our narrative)

Similarly the working of *any* organ or *any* body part can happen only after the action cycle as described in the right eye dust particle example is completed!

In this sense the brain *represents the whole body!*

The brain and other body parts of humans is the result of biological evolution; they can function only if life is infused into them. This is provided by the mind.

Swami Krishnananda Saraswati (1922-2001) General Secretary of the Divine Life Society, Rishikesh, India has this to say:

*The mind is the consciousness of individuality, and it is present in every created thing.......*

*Essentially, the whole mind is cosmic......It appears like drops, or waves, or ripples operating in different persons, things, individualities etc*

On its relationship with the brain the Swami goes on to say:

*You may say the brain is made of material stuff. Is the mind also made of material stuff?......*

*If it is made of the same stuff as the brain it cannot be **conscious**.....*

*So we must say that mind is made up of **non matter...***

*If the mind and brain were two different things, you would be feeling like a **split personality**....*

*You cannot be made of two things and the feel that you are one.....*

*Actually, the brain is a solidified form of the mind itself; as water looks like ice, the mind looks like the brain, and the body.*

Swami Krishnanda with the above explanation has tried to make an abstract topic intelligible to the layperson!

This essay has attempted to probe in depth the working of the *material* human brain and the *immaterial and intangible* human mind with which it is traditionally linked. For most of us the brain and mind are the same, and in common parlance they are used interchangeably. This is incorrect. The brain is matter and the mind is definitely *not* matter. In day to day living the brain and mind do overlap. As we have seen, It is not easy to identify the difference in a manner which is acceptable to the scientist. The philosopher or spiritualist may accept the reasoning and the conclusions arising from it.

Meanwhile further researches by neurologists on the less understood workings of the human brain at the microscopic level continue.

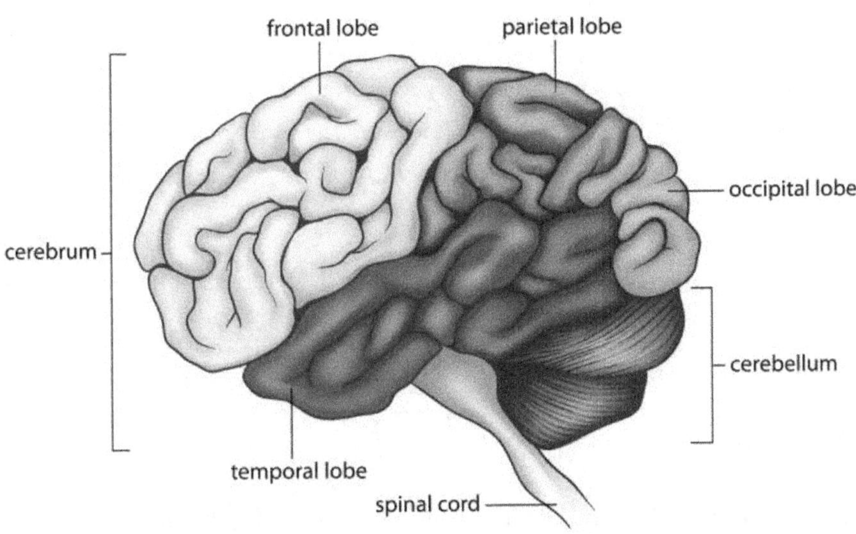


**Illustration 8:** Human Brain Anatomy- Shutterstock _ 139691548

# Nine

## Kidneys ~ Master Chemists of The Human Body

*Every 24 hours, the kidneys sweep clean of wastes over a TON AND A HALF of blood (as the body's supply circulates and recirculates). And as a built-in safety feature they have well over twice the capacity needed to maintain health: thus if it is necessary to remove a diseased kidney, the remaining healthy one does double duty with ease.*

**J.D.Ratcliff (1914-Unknown)**

## FUNCTION:

The kidneys would rival the liver as the body's *master chemists*. They regulate and fine tune the water and minerals like potassium, sodium and chlorides in the blood within a definite range that are vital for human survival e.g. a high level of potassium could stop the heart as effectively as a bullet passing through it.

The kidneys could similarly be compared to another important organ, the heart in respect of dimensions and weight. Averages relating to males are given below:

HEART

Length(cms) 12

Width(cms) 8

Thickness(cms) 6

Weight(gms) 298


KIDNEY

Length(cms) 11.2

Width(cms) 5.6

Thickness(cms) 3.8

Weight(gms) 150


Considering that there are TWO functional kidneys, a SINGLE heart and both organs have continuous non stop flow of blood through them, the comparison is revealing!

Another interesting statistic relates to the measurement of the blood flow in men expressed in ml/minute/100 gms of organ weight:


KIDNEYS 309

LIVER 96

HEART 70

BRAIN 54

This comparison shows that the blood flow every minute per 100 gms of organ weight in the kidneys is more than the liver, heart and brain PUT TOGETHER!

Aside from this comparison with major organs of the body, the kidneys have been correctly described as the 'body's sewage treatment plant'!

This is by far its most important function!

The kidneys flush out urea, uric acid creatinine, and other ammonia compounds out of the system. Other organs perform similar functions:

Through its sweat glands, our skin periodically removes sweat (urea and minerals) the quantity depending on climatic conditions, physical activity and stress level.

The lungs expel carbon dioxide through respiration.

The liver and intestines remove solid body waste as part of the digestion process.

A comparitive study of these excretory processes shows that In terms of a healthy body, as related to the flushing out of harmful substances, the kidney function is right on top.

## STRUCTURE:

The kidney has a master plumbing architecture that would rival the sewage treatment and disposal systems of a major city!

## IMPORTANCE:

In an earlier post we had debated the nomination of the first and second positions for body organs in terms of their importance. We had awarded the first and second positions to the heart and brain respectively!

If a third position is to be nominated I would vote for the kidneys!

Some may be of the view that such comparison of body organs is meaningless

After all, The working of the human body is similar to that of a big city- if there is a major traffic pile up on one of the highways, the other roads leading into and out of the city will be subjected to traffic congestion and delays.

Likewise the human bodies main organs are closely inter related. Malfunctioning of one will impact the others very quickly.

Nobody can deny this fundamental truth; it is however important to understand the functioning of the main organs as a guideline for a healthy life.

Why are the kidneys so important?

Let us probe further into their structure and the work they do day in and day out!

There are 2 kidneys, right and left located at the rear of and towards the base of the chest cavity or thorax. The right kidney is below the liver and the left below the spleen. The liver being the largest organ in the body tends to depress the position of the right kidney; the spleen being a smaller organ has a lower depressing effect on the left kidney.

For this reason, the right and left kidneys are not in a straight line.

They are protected from external injury by the 11th and 12th ribs of the ribcage.

The left kidney is more centred in respect of the sternum or breastbone as compared with the right.

Although the functions of the two kidneys are identical, the left kidney is slightly bigger than the right.

A woman's kidneys are slightly smaller and weigh slightly less than a man's but are identical in respect of structure and function.

The intricacy of the structure and complex working of the kidney is mind boggling!

Each kidney is made up of more than a million microscopic nephrons.

Each nephron has a relatively large head called the 'glomerulus' and a thin convoluted tail called the 'renal tubule'

The glomerulus is covered by a web of tiny capillaries (one tenth of a mm in thickness) that filter the oxygenated blood supplied to it.

The blood flow to the kidneys commences with the pumping action of the left ventricle of the heart through the aorta and an arterial network spanning the entire body from crown to toe.

Within the kidney, approx 1700 litres of blood flows every day. They get 'purified' i.e.made free of toxins like urea, uric acid in the intricate web structure of the glomerulus. The purified filtrate consisting mainly of water and substances like amino acids is referred to as 'primary urine'.

About 170 litres of 'primary urine' is produced each day. Since they contain water and nutrients required by the body they go through a process called 'reabsorption'.

Following reabsorption, the volume of residual fluid constitues the 'urine' we are familiar with. This finds its way through the 2 ureters (1tube from each kidney) to the bladder where it is stored and regularly voided through urination.

The kidney does not have a pumping action like the heart for moving the materials within it from time to time.

The processes that enable these changes are 'diffusion' or 'osmosis'.

The working of the glomerulus and tubule is not a chain of repeating cycles.

The nephron in the course of its working 'develops' i.e. its form and function varies depending on the requirement of the kidney as an organ from moment to moment!

This is one of the amazing truths of the kidneys that renders it unique!

As an example of this change of structure, the nephron at a particular stage of its development resembles the tadpole like male sperm that fertilises the female ovum to form the smallest unit of 'created life' called the 'zygote'

At a later stage of its functioning the size

and shape of the glomerulus changes and the renal tubule becomes more convoluted!

Are these series of precisely timed actions a result of design by an omniscient Creator or the outcome of evolution over aeons of time?

This quandary applies not just to the kidneys but generally to each and every organ in the human body!

The fine tuning and coordination in major organs in the body like the kidneys is truly incredible!

## HUMAN KIDNEY ANATOMY

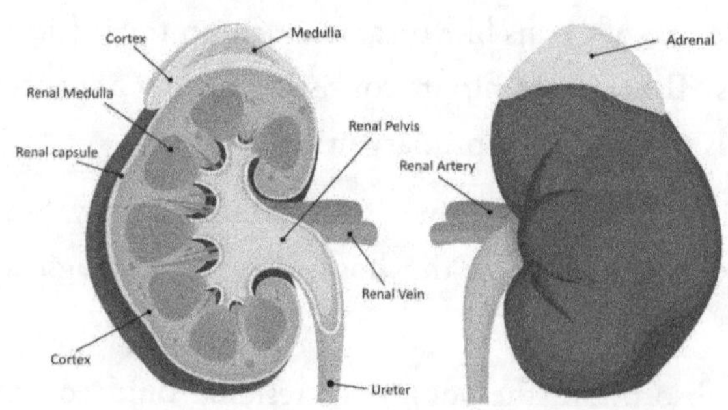


**Illustration 9:** Human Kidney Anatomy- Shutterstock _1316280749

# Ten

# The Endocrine Glands

*Natural forces within us are the true healers of disease.*

**Hippocrates Ancient Greek physician**

This essay relates to the endocrine glands. For a proper perspective we shall present an overview of the human body, more specifically the human brain which controls the functioning of such glands.

I shall briefly digress:

There are a dozen systems within our bodies which function day and night *without a break* literally from womb to tomb. The systems are controlled by the human brain. Their proper functioning bestows the gift of *life*. When the systems cease to function, *death* of the body ensues.

The complex command and control centre in the brain enables the *fantastic* mental capabilities of humans e.g. the human brain can outperform the the world's fastest supercomputer. As per a recent study, the Chinese *Tianhe-2,* the fastest supercomputer with a maximum processing speed of 54.902 petaFLOPS (one thousand trillion calculations per second) falls way short of the processing speed of the human brain! This truth is not self evident. We need to realize what the human brain as a whole is capable of doing:

*It links interconnected areas through billions of neurons and possibly trillions of glial cells which enables it to perceive,interpret,store,analyze, and redistribute at the same time!*

No computer can match this all round wizardry of the human brain!

The dozen systems mentioned earlier such as the *circulatory, respiratory, digestive, nervous, reproductive,endocrine* consist mainly of organs, glands, ducts, blood, and blood vessels. We shall not get into the details of each since they are not relevant to our narrative.

Instead, we focus on one of them I.e.the *endocrine system*, more specifically on the endocrine glands.

An interesting feature of the endocrine glands, unlike other glands in the body is that they are *ductless* I.e. their secretions called *hormones* are released directly into the blood stream and can speedily reach any part of the body whenever required.

In contrast,other gland secretions called *enzymes* are delivered through ducts and have a *local* effect.

This difference is exemplified by:

1) The two adrenal glands, atop each kidney are typical endocrine glands that secrete the hormone *adrenalin.* When an acute danger or threat is perceived additional adrenalin is secreted and released into the bloodstream, generating extraordinary physical and mental capabilities. The concerned individual faces a *fight or flight* option, makes a decision and acts *all in a matter of a few seconds* e.g. at great risk to himself, he rushes into a burning aircraft to save hapless trapped passengers.

2) Compare this with the innocuous exocrine salivary gland duct that delivers the enzyme amylase to convert starch to maltose in the mouth as a part of the routine digestion process!

Hormones and enzymes are complex organic chemicals secreted in the human body in miniscule quantities; there are 50 known hormones and a possible 75000 enzymes (some still under investigation) engaged in a wide spectrum of chemical reactions from crown to toe!

The important functions they perform are way out of proportion to the small quantities produced -*extremely vital* for the body's well being.

To get a better idea, imagine a road repair project being executed with a team of workers and a highly qualified engineer supervisor. For a fair assessment of relative importance, hormones can be likened to the *engineer supervisor* and enzymes to the *workers* who carry out orders; in the absence of the supervisor, the workers will do *nothing*; to borrow an expression from Hindi,the supervisor/worker nexus resembles a *guru/chela* relationship.

There are 15 glands in the human body- some exocrine with ducts, some endocrine without ducts, some having dual functions,both exocrine and endocrine. Examples:

## Exocrine:

1. Pancreas-secretes *enzymes* amylase,trypsin,and lipase which digest carbohydrates,proteins, and fats respectively.

2. Sweat glands-secrete sweat for *homeostasis* (regulation of equilibrium factors within the body e.g. body temperature at 98.4 deg F irrespective of the outside temperature)

3. Sperm-secretes *enzyme* hyaluronidase that jogs the male sperm along in its arduous journey to unite with the ova.

## Endocrine:

Secretions are hormones. Main endocrine glands are:

1. Anterior and Posterior Pituatary glands- These are located below another endocrine gland the hypothalamus in the brain. The duo work in tandem.

   The Anterior Pituatary secretes the hormone prolactin (stimulates milk production in mothers), supplements the thyroid gland production of T 3 & T 4 thyroid stimulating hormones, supplements production of hormones related to the adrenal glands (see details under adrenal glands)

   The Posterior Pituatary secretes an anti diuretic hormone which conserves body water by minimizing water loss from kidneys. It also secretes oxytocin which in females stimulates contractions in the uterus after full gestation.

2.  Thyroid gland- T 3 & T 4 thyroid hormones.

3.  Adrenal gland- secretes gluco-corticoids in the cortex part of the gland and nor-adrenalin in the medulla part.

There are five more endocrine glands.

We shall provide more details on the endocrine system later in the narrative.

## Exocrine and Endocrine (Heterocrine glands):

1.  Pancreas:

    Exocrine functions: secretes enzymes like amylase.

    Endocrine functions: secretes hormones like insulin.

2.  Liver:

    Exocrine functions: secretes bile into the intestine through the bile duct.

    Endocrine functions: secretes the hormone Insulin like Growth Factor1 (IGF-1)

3.  Kidneys:

    Exocrine functions: secretes urine

    Endocrine functions: secretes hormones like calcitriol and renin

4.  The male and female sex organs, *testes* and *ovaries*:

    Exocrine functions: Reproductive system is functional through secretion of sperm (male testes) and ova (female ovaries)

    Endocrine function: Male testes secretes hormone *testosterone* the origin of male libido; female secretes hormone *estrogen* that triggers ovulation.

The "The Amazing Endocrine Glands" is the title of this essay. What is special about these glands that they stand out from the rest?

A partial answer has already been given in the hormone/enzyme comparison I.e.supervisor giving orders/worker carrying out the orders.

To throw more light on the topic, let me start with a quote:

*Three mysterious substances,in minute quantities,control the chemistry of the human body. Some of the most important facts about them have been unearthed only within recent years. These substances are the* **hormones***,those powerful chemicals secreted in the body by the endocrine glands; the* **enzymes***, which turn one chemical substance into another; and the* **vitamins***. These magic chemicals maintain an extraordinary balance among forces so powerful that any of them could be destructive if unchecked.*

### Bruce Bliven, Jr (1916-2002) American author

Hormones secreted by the endocrine glands have been defined as *chemical messengers*. The definition fails to do full justice to their role in ensuring normal development of vital organs of the human body- *magical chemical substances* would perhaps be more appropriate.

Physiologist J.D.Ratcliff (1914-Unknown) defines them as:

*Hormones circulating in your bloodstream are much like miniature hydrogen bombs - powerful almost beyond belief.*

Definitions apart, following facts highlight the preeminent role played by hormones in the functioning of the human body:

1.  During her 30 child bearing years a woman's ovaries secrete approx 100 milligrams of the hormone *estrogen*, equivalent in weight to 50 mustard seeds.

    As a growing child she reaches puberty, and an amazing metamorphosis takes place-the equivalent in weight of less than one mustard seed of this hormone transforms her body to that of a woman!

2.  Magical effects can be observed in the thyroid gland which secretes the T 3& T 4 hormones-even a nominal below par production in the mother during the gestation period can transform a newborn baby to an unfortunate malformed idiot!

3.  The pituatary and hypothalamus endocrine glands, arguably the most important endocrine glands at the brain base secrete the growth hormone *somatotropin*. A fluctuation in normal output leads to serious

consequences for the body- the gigantic circus strongman has too much of this hormone, the dimunitive dwarf too little!

We have in our narrative so far, referred to hormones as applicable to humans. They in fact apply in equal measure to animals e.g. your favourite film star's sex hormones cannot be distinguished from those of a lion. Hormones from your favourite cricketer's pituatary gland is no different from those of a mouse!

A common factor in hormones,enzymes, and vitamins is that they cause amazing changes in *the individual miniscule body cell,* (the human body consists of trillions of such cells) following a *precise timetable* in ways that are still not fully understood! This activity is referred to as *metabolism* and each of these magic chemicals produce their individual *metabolites* that characterize the changes!

It would be fair to designate the hormone produced by the endocrine gland, as leader of the three member pack in ensuring timely changes in development of the human body and carrying out other vital functions.

We now focus on *vitamins* possibly having functions similar to hormones; such links are under active investigation by molecular biologists.

(In common parlance vitamins are associated with food supplements that compensate for dietary deficiencies in humans particularly with advancing age)

Let us look at the function of vitamins as a whole with respect to the human body:

*A vitamin is a substance that makes you ill if you don›t eat it.*

### Albert Szent Gyorgyi -Nobel Prize in Physiology or Medicine 1937

To elaborate on Gyorgi's statement, vitamins is an abbreviation for *vital amines*; along with their metabolites (products of metabolism) they fulfill a wide range of physiological functions such as hormones and antioxidants, as regulators of tissue growth, in the development of the human embryo and many other activities. They also play a role in the immune system.

Vitamin A has historically been known as the *anti infective* vitamin and Vitamin D the *anti ricketic vitamin.* These definitions are incomplete and outdated. Recent studies have shown that their activity extends way beyond such facile descriptions.

Vitamins A and D bio active metabolites, retinoic acid and the D 3 molecule have hormone like properties that influence the immune system in highly specific ways; besides they modulate

vital immune responses like lymphocyte activation and proliferation when required.

Vitamin C and E and constituents of Vitamin B Complex act in a relatively non specific manner e.g. as antioxidants.

In conclusion, we can say that hormones,enzymes,and vitamins play a major role by enabling timely complex chemical changes in the cell structure of organs in the human body associated with growth. They are integral parts of the body immuno system. Good health and life itself would not be possible without them. The precise nature of the chemical reactions is complex organic chemistry which we are not getting into.

We had nominated hormones, produced by the endocrine glands as the leader of this formidable triad!

Questions remain: What *prompts* hormones to trigger the multifarious chemical reactions in the human body as per such a *precise timetable?*

What *enables* enzymes to carry out the chemical reactions with such *precision?*

What *enables* vitamins to perform their *specific* and *non specific tasks* so effectively?

Science does not as yet have convincing answers!

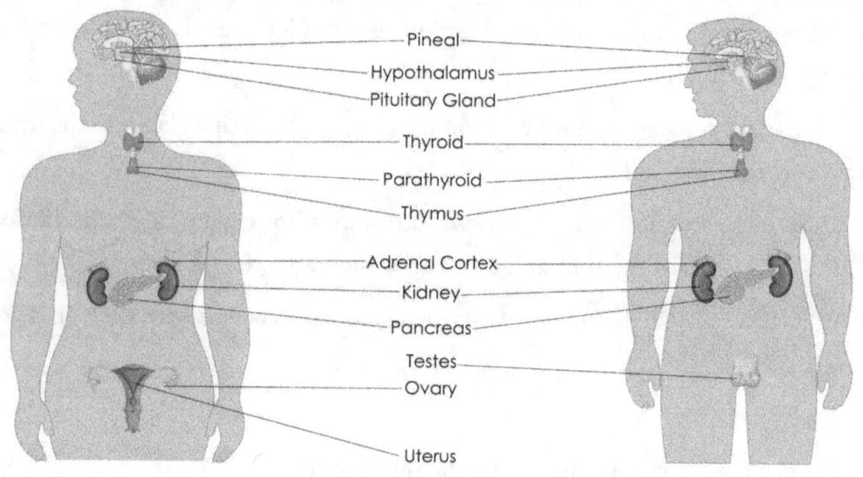


**Illustration 10:** Endocrine System Male - Female- Shutterstock _1057812977

# Eleven

# Vitamin B 17 ~ The Cancer Connection

## CANCER AND VITAMIN B17

*Cancer is a marathon - you can't look at the finish line. You take it moment by moment, sometimes breath by breath, other times step by step.*

**Sarah Betz Bucciero American Author**

Worldwide, an estimated 19.3 million new cancer cases and almost 10.0 million cancer deaths occurred in 2020. Of these, female breast cancer is in the top spot surpassing lung cancer as the most commonly diagnosed cancer with an estimated 2.3 million deaths. In India, the projected number of cancer patients among males was just under 7 lakhs and in females slightly exceeding 7 lakhs. It is forecasted that 1 in 9 Indians will develop cancer during their lifetime (0 - 74 years)

Related statistics indicate that one woman dies of cervical cancer every 8 minutes in India.

Also, in India:

1. For every 2 women newly diagnosed with breast cancer, one woman dies of it.

2. Cancer deaths due to various forms of tobacco abuse is estimated to exceed 3500 persons every day, or about 1.3 million per year. Of these, Tobacco (active and passive smoking) accounted for over 3 lakh deaths of men and women in 2018.

3. The highest number of cancer cases are in Kerala, followed by Mizoram, Haryana, Delhi and Karnataka.

The above numbers clearly indicate that cancer is a major health hazard in the world including India and in need of an intense probe to cover causes, effective diagnosis, and possible treatments,

Cancer comes from overproduction and malfunction of the body's own cells. Anything that may cause a normal body cell to develop abnormally can potentially cause cancer.

There are over 200 types of cancer. The main causes as stated by the scientific community are:

1. Exposure to chemical or toxic compounds.

   Examples are asbestos, tobacco or cigarette smoke, benzene, nickel, cadmium.

2. Ionizing radiation.

   Like uranium, ultraviolet rays from sunlight, X rays, Gamma rays.

3. Pathogens - Hepatitis viruses B and C, HPV (Human papilloma virus), EBV (Epstein Barr virus) amongst others.

   Several types of bacteria are also being investigated.

4. Human genetics.

   Common examples are breast, ovarian, colorectal, prostate, melanomas starting from the skin caused by excessive cells of the pigment melanin and gradually spreading mainly to the eye, brain, and lymph nodes.

Cancer symptoms and signs depend on the specific type and grade of cancer. Some of the common symptoms are fatigue, weight loss, pain, skin changes, change in bowel or bladder function, unusual bleeding, persistent cough or voice change, fever, lumps, or tissue masses. Definitive diagnosis is made by examination of a biopsy sample of suspected cancer tissue.

Cancer 'staging' helps to determine the cancer type, the extent of cancer spread and to fix treatment protocols. The aggressiveness of the cancer is established on a scale 0 to 4, representing the lower and upper ends of the spectrum.

Treatment protocols usually include at least one if not all of the following:

Surgery, chemotherapy, and radiation therapy.

Prognosis of cancer cases, particularly falling in the categories 3 and 4, despite advances in medical science are still considered to be tentative.

This uncertainty has introduced an air of mystery regarding the cause of the disease, it's diagnosis and treatment.

Despite the tremendous advancement in medical sciences in the 21st century, it is faced with nagging questions for which we still do not have precise answers:

What actually causes cancer?

What are the available treatment options?

Is the outcome in majority of the cases treated by existing methods a partial or a complete cure?

In this context, the following ideas of Dr. James Liu MD, Specialist in Reproductive Endocrinology, OB/GYN - Gynecology from the UH Cleveland Medical Center, Cleveland, Ohio are relevant:

1.  Most people believe that cancer is like a flu, a cold, something we can shake off. It is not so!

2.  Cancer to put it plainly is the body's "failure to keep consistent DNA" Our DNA and protein structures replicate at an alarming rate. Billions of cells replicate to grow us into adulthood and beyond. When replication occurs, there is a very very small chance that the replication code will have an error like

    ACGTACGT (Acronym for Adenine, Cytosine, Guanine, and Thymine) becoming ACGTTCGT.

In most such cases, benign errors like replication on skin cells do not have an impact. However, some specific genes are critical for cell growth and cell death.

When the normal replication code, "grow every 5 days, die after 5 days" becomes "grow every 5 days, die after 15 days we call that "cancer".

Likewise, if the code changes to "grow every 5 days die after 2 days" there is an imbalance in the replication equation leading to a "cancer" type condition in the body.

In either case there are "more cells growing than are dying"

What is more, for the survival of the excess cells created by the cancer it steals vital nutrients from other critical organs that over a period of time can become life threatening!

3.  Science is struggling to find viable treatment options for cancer. No fool proof method has been discovered so far. A major problem with some of the options like "chemotherapy " is that while providing relief by attacking cancerous cells, the treatment also causes harm to the neighbouring healthy cells.

We can summarize the available treatment options as follows:

a)  Cut out the cancerous growth.

b)  Burn it with radiation.

c)  Freeze it to death

d)  Target it with chemicals.

e)  Strengthen immune system.

f)  Replace patient's diseased organ with a compatible healthy donor organ.

Science is simultaneously working on the following refinements to existing treatment options:

1.  Early diagnosis

2.  Advanced surgical methods for removal of cancerous growth.

3.  Superior Target Chemicals

4.  Hack and hijack the patient's immune system to do our bidding.

Another aspect of the cancer treatment story is the progress in the development of new drugs in the management of the disease.

In the U.K. in 1930, a herbal preparation named "Vinculin" was marketed as a possible alternative for diabetes patients who had developed a resistance to insulin. It met with limited success.

In 1950, similar research was conducted by 2 teams from the US Pharmaceutical giant Eli Lily and the University of Western Ontario on the "Madagascar periwinkle" a herbaceous plant that grew in subtropical regions like Australia, Malaysia, India, Pakistan and Bangladesh yielding pretty flowers varying in colour from white to dark pink with a darker red centre.

The plant extracts were found not very effective in lowering blood sugar but they dramatically decreased the number of white cells in the blood. Since some types of cancer such as leukemia involve a proliferation of white blood cells the researchers quickly realized that Madagascar periwinkle extracts could contain chemicals with potential cancer fighting properties.

Their collated findings led to the rapid identification, development and advancement of 2 closely related drugs from the periwinkle extract, "Vincristine" and "Vinblastine"

Chemically, VC and VB are similar alkaloids.

VC is used in the treatment of children's leukemia. VB is used to treat "Hodgkin's disease", a form of lymphoid cancer.

In this context, the work done by American molecular biologist Professor Sarah E. O'Connor is noteworthy. She is investigating the whole sequence of metabolic events that the Madagascar periwinkle deploys to produce VC and VB. Her research has identified key genes involved in their development but a few are still missing. Once the pathway is completed, it will be possible to engineer other organisms to increase the world supply of these valuable chemicals.

Hopefully, her efforts will enable tapping of the full potential of this remarkable plant.

Madagascar periwinkle to people living in subtropical climes is just a 'weed' and to the western gardener an easy to grow 'living room ornamental' till a century back. Today it is the starting point for life saving drugs!

Of the other available treatment options, the "Cytotron" Technology is noteworthy.

Elderly patients in India unwilling to undergo conventional treatments have reported encouraging results after a month of RFQMR (Rotational Field Quantum Magnetic Resonance) therapy).

Dr.Rajah Vijay Kumar, Chairman, "Oganization de Scalene" is the inventor of the tissue engineering technology of RFQMR therapy and treatment options using the Cytotron device.

Dr. G.S.Nayar in Bengaluru a cofounder of "Ojus Health Care" is considered by many as another pioneer in this field. The Cytotron is probably the first device of its type to be fully conceived and developed in India and accredited internationally.

A revolutionary theory having its origins as far back as the 19th century suggests that cancer is not caused by uncontrollable gene mutation, as commonly believed but by the deficiency of a member of the extended Vitamin B Complex molecule called Vitamin B17. In other words it is not a result of human genetic abnormalities or an external agent but by a deficiency in the nutrient inputs of humans in different parts of the world.

The protagonists of this theory point out that the dreaded disease scurvy, scourge of maritime seafarers from Europe in the 17th century was eliminated through large doses of Vitamin C. Vitamin C was administered through daily intake of oranges and lemons that ensured an increase in ascorbic acid (Vitamin C) level in the patient's blood. Scurvy was characterized by 'fungous flesh, putrid gums, fever and red patches in many parts of the body. Untreated, many sailors suffered prolonged painful illness culminating in death and burial in the sea!

There are more instances of Vitamin deficiency leading to different types of diseases: Deficiency of Vitamin A makes the patient prone to infections. Vitamin A is referred to as the 'Anti Infective vitamin.'

Deficiency of Vitamin B makes the patient prone to 'beriberi' (characterized by fast heart rate, and numbness in the hands and feet) Vitamin B is often referred to as the 'Anti beriberi' vitamin.

Vitamin D deficiency is known to cause rickets (abnormal bone formation) in children and referred to as the 'Anti Rickets' vitamin.

We need to critically examine these claims:

The Vitamin B Complex molecule has a known range from B1 to B12.

Vitamin B17 is still in the exploratory stage. It is not even certain whether it is a vitamin at all since it does not exhibit properties similar to the established water soluble Vitamins B1 - B12, and C,or the fat soluble Vitamins A,D,E and K (13 in all)

At this stage we need to introduce a class of chemicals called 'Phytochemicals' in our narrative.

Phytochemicals are a wide variety of non nutritive chemical compounds found in plant foods with possible health benefits. They protect cells and DNA from damage that may lead to cancer.

There are three major groups of phytochemicals:

Polyphenols- sub categorized as the flavonoids, phenolic acids and other non flavonoid polyphenols.

Terpenoids- sub categorized as the carotenoids (like beta carotene in carrots, lycopene in tomatoes) and non carotenoid terpenoids.

Thiols- glucosinolates, allylic sulfides and non sulfur containing indoles.

Scientists estimated there are more than 5000 phytochemicals. Studies show that they can help keep cancer causing agents from forming.

We focus on the phytochemical "Laetrile" (a possible candidate for Vitamin B17)

Claims that laetrile or amygdalin can treat cancer do not as yet have sufficient research back up. Further, these products are known to contain cyanide that can cause serious side effects.

Laetrile is a partly man made (synthetic) form of the natural substance amygdalin. Amygdalin itself is a plant substance found in raw nuts, bitter a almonds, as well as apricot and cherry seeds. Plants like lima beans, clover and sorghum also contain amygdalin.

Laetrile has been used as an anticancer agent since 1800's. It is used either on its own or as part of a programme. This might include following a particular diet, high dose vitamin supplements and pancreatic enzymes.

Although, more recent studies have shown that laetrile or amygdalin can kill cancer cells in certain cancer types there is not sufficient scientific evidence to show that laetrile or amygdalin can treat cancer Despite this, it still gets promoted as an alternative cancer treatment.

What exactly are the sources for Vitamin B17 and is the comparison of cancer in the 21st century with scurvy in the 17th century a case of 'Sauce for the gander is also sauce for the goose'?

In this context consider the following quote:

'Chemotherapy and radiotherapy will make the ancient method of drilling holes in a patient's head to permit the escape of demons look relatively advanced.....

*"Toxic chemotherapy is a hoax. The doctors who use it are guilty of premeditated murder, and the use of cobalt and other methods of cancer treatment popular today effectively closes the door on cure."*

[Dr.Ernst T.Krebs (1911-1996) American biochemist]

Strong language indeed considering the large number of surgeries, chemotherapy and radiation treatments being carried out in hospitals the world over on cancer patients!

Vitamin B17 is also marketed in synthetic capsule form. Protagonists of Vitamin B17 capsules claim it possesses prophylactic and curative properties for cancer.As of now there is no conclusive evidence that this claim is justified.

Considering the strongly invasive nature of surgery, chemotherapy, and radiation therapies, low success rate and the debilitating side effects vividly described by Dr. Krebs it is worth giving Vitamin B17 a try as a prophylactic and/or cure for certain types of cancer for a limited period of time.

It does seem it can do no harm.

# SPECIAL FEATURES OF THE HUMAN BODY

# Twelve

## The Third Eye

*I do not believe that God ever intended to disclose to man what man could find out for himself.*

**Georges Lemaitre, Belgian Catholic Priest, Astronomer and Physicist**

The Hindu Trilogy, cornerstone of Hindu religious belief commences with BRAHMA-THE CREATOR, followed by VISHNU-THE PRESERVER and ends with SHIVA-THE DESTROYER.

The Trilogy symbolically represents the relentless unending cycle of our universe.

The origin of the Trilogy is shrouded in mystery. Most Hindus believe that it dates back to the Rigveda circa 1500 B.C.E.

Be that as it may, our attention focuses on SHIVA-THE DESTROYER more particularly his *third eye* which is relevant to our narrative

Images of Shiva show the third eye in a dormant state as a vertical line on the forehead; when this state of mental equilibrium is violently disturbed, the vertical line eye is believed to blink and open into a full fledged eye, blazing and fearsome.

As per ancient folklore, myth, and Hindu belief there are only a limited number of occasions when such an event has actually occurred. We shall not get into these details since they are not needed for the broad message our narrative seeks to convey.

The third eye attributed to Shiva and as briefly described above is possibly present in each one of us but dormant most of the time..

Throughout the evolution history of man,Neanderthal onwards there is no recorded instance by paleontologists of the sighting of a three eyed human ancestor. Exception is in legend like the savage Cyclops with a single mid forehead eye in Greek and Roman mythology circa 1200 B.C.E.

Aside from this, the third eye though mysterious and symbolic, has struck a chord in the mind of the investigative physiologist,spiritual leader and occultist:

1) Does it have a connection with the Hindu concept of *kundalini* or with the mysterious *seven chakras* which are believed to exist within the human body?

2) The physiologist would ponder on possible links with the endocrine pineal/ pituitary glands or the lymphatic system. The researcher in the field of occultism would query whether the third eye is a manifestation of ESP?

There is a plethora of possibilities. In the course of our present narrative we shall try to find a cogent link between them, that may lead to acceptable conclusions.

1.  KUNDALINI/CHAKRAS: As per Hindu Vedic scripture the seven chakras represent seven *energy points* in the human body through which *the prana* or life force flows.

Kundalini represents as per Hindu belief, two coiled serpents closely intertwined in three and a half rolls at the base of the human spine resembling a DNA double helix. The serpents are in a state of deep slumber for most of an individual's life. The three rolls refer to the conscious, subconscious, and unconscious states of an individual's life and the last half to the *turiya state* attained by sages and mystics following a lifetime of intense yogic practices. The state of deep slumber refers to the individual's conscious state I.e. the awakened state in the material world. This state relates to the fulfillment of sensual/material desires and is totally oblivious of the much larger and deeper paranormal and spiritual states represented by the subconscious and unconscious, beyond the human intellect.

For sincere seekers of the paranormal and spiritual the kundalini triggers the gradual uncoiling of the serpents from the base of the spine i.e.

from the *mooladhara* or first chakra upwards to the pinnacle position of *sahasrar* identified with the sixth and seventh chakras located midpoint of the base of the two brain hemispheres. This is another way of saying that the concerned individual travels from the material conscious world to the subconscious and unconscious i.e. paranormal and spiritual worlds.

In exceptional cases he even reaches the turiya state of pure bliss.

(In the Hindi speaking regions of modern India, where most Hindus live, a related word *kundali* is often used which has a different meaning I.e. horoscope or birth chart)

Nobody has *seen* any of the seven chakras awakened by the kundalini in a X Ray, MRI or any other picture of the entire length of the human spinal cord.. It is logical to conclude that they are nonexistent in a physical sense. They represent symbols of ancient Hindu arcane concepts linked with yogic practices; their credibility rests in the supernatural powers exhibited by yoga practitioners who have successfully activated their kundalini/chakras over several centuries right upto the present day e.g. Sage Agastya (1500-1200 B.C..E.), Sages Vasishta, Vishwamitra, and Valmiki (mentioned in the Rigveda), and Patanjali in 200 B.C.E. In modern times we could include Ramana Mahirshi, (1879-1950) Saibaba of Shirdi, (1835-1918),Osho Rajneesh (1931-1990) and many other mystics in Asia and the western world like Americans Edgar Cayce (1877-1945) and Claudia Brownlie. These mystics exhibited superhuman powers indicative of awakening of their respective kundalini/chakras.

2.  THIRD EYE: There is a definite correlation between the chakras and the third eye,more specifically the sixth and seventh chakras whose location in the human brain is suggestive of a link with the endocrine pineal and pituatary glands respectively.

The third eye also is believed to be connected with the paranormal field, more specifically with ESP.

To clarify and validate these statements I would need to digress- We rewind our evolution clock by a massive 300 million years to the dinosaur period. This can be described as the age of reptiles I.e. cold blooded creatures.

Apart from variants in the dinosaur family, there were crocodiles, the snake,lizard, and many other reptile species who possessed the third eye in the form of the cone shaped pineal gland located in the brain behind the two visible eyes. In the evolution process extending to the warm blooded mammal era commencing 250 million years back, the pineal gland seems to have atrophied in the anatomy. A possible explanation is that with the development of sensory perception through the normal two eyes, the need for the intuitive third eye gradually reduced and finally faded away.

This could explain how the ancestors of primitive man who is a warm blooded mammalian vertebrate lost the pineal gland and the third eye for millions of years until 1958 when its presence in humans was once again established. In contrast the presence of the pituatary gland located just below the pineal in the human brain has been known for the last three centuries.

An interesting example of the presence of the pineal in cold blooded reptiles is the fearsome giant lizard *Komodo Dragon* in the Indonesian islands. This sluggish predator seems to rely on its paranormal powers of the pineal gland to trap its prey.

For humans of the present day, it is reasonable to conclude that the pineal gland exists, which when activated releases paranormal power,not as dramatic as Shiva's third eye where there is a visible opening of the third eye, but rather as an illuminating flash unexplainable as a sensory perception..

The multitasking duo of the pituatary gland and hypothalamus one of the most important endocrine gland systems in humans controls hormones of two other endocrine glands,the thyroid and adrenals; the thyroid is located below the Adam's apple in the throat. The two adrenals lie atop the two kidneys. We shall not get into these details except to comment that the pituatary gland linked to the *seventh chakra* does not appear to be the third eye associated with paranormal powers.

In view of the intriguing role played by the pea sized pineal, we shall explore this mysterious gland further:

The pineal gland derives its name from the coniferous pine tree one of the earliest vegetation forms on earth. Apart from the organ itself, there are spiritual connections.

In images showing the crown of the Buddha and Shiva himself the pine cone is clearly seen. In the case of Shiva, additionally, there is the presence of the coiled serpents related to the third eye. Philosophers and the bible (in both the old and new testaments) also refer to the third eye.

French philosopher Rene Descartes describes the pineal gland as *the seat of the soul.* Indian Saint Paramahamsa Yogananda offers the following explanation for verses 1-2 of Ezekiel 43:

*Through the divine eye in the forehead* (the east) *the yogi sails his consciousness into omnipresence, hearing the word or Aum, the divine sound of* (many waters) *the vibrations of light that constitute the sole reality of creation.*

In the new testament, Jesus says:

*If therefore thine eye be single, thy whole body shall be full of life.* (Matthew 6:22)

Some interesting sidelights on the pineal gland:

1. The pineal gland secretes the neurohormone melatonin when we are asleep mostly at night in complete darkness. This centuries long connection between melatonin and darkness has a spin off.

   When you are unable to sleep e.g. during a long intercontinental flight, a melatonin tablet could induce a feeling of darkness with a good chance of falling asleep. It is a method of taking advantage of the body's circadian rhythm.

2. The awakening of the kundalini or third eye/pineal gland is associated with the release of the psychedelic drug dimethyltryptamine in the human body; in fact during any intense paranormal or spiritual experience like a near death experience when the pineal gland is activated DMT is released.

3. The tiny pineal gland, because of its extraordinary versatility, is serviced by a volume of blood supply next only to the kidneys.

4.  For those interested in mathematics, the shape of pineal gland I.e. pinecone is a perfect example of the *Fibonacci spiral* I.e. a structure whose supporting columns form a Fibonacci sequence (0,1,1,2,3,5,8,13,21,34, 55....... is a Fibonacci sequence

where a number in the sequence is the sum total of the preceding two numbers). If we form squares of the Fibonacci numbers, in ascending order I.e.small to large, and connect the bottom corner of the smallest square to the opposite corner of the next square and take the process forward you get a beautiful Fibonacci spiral which will fit exactly into the shape of an ascending pine cone!

In the narrative so far we have spoken of ancient traditions and religious beliefs which have stood the test of time and are observed even today. They are articles of faith handed down over centuries and the number of adherents keeps increasing.

Hardly any of the claims would hold up to modern scientific tests; they are at best pseudosciences, unacceptable to the scientific purist.

Will we ever be able to prove that some/all of them are *true* or *false*, only time can tell. As of today we do not know.

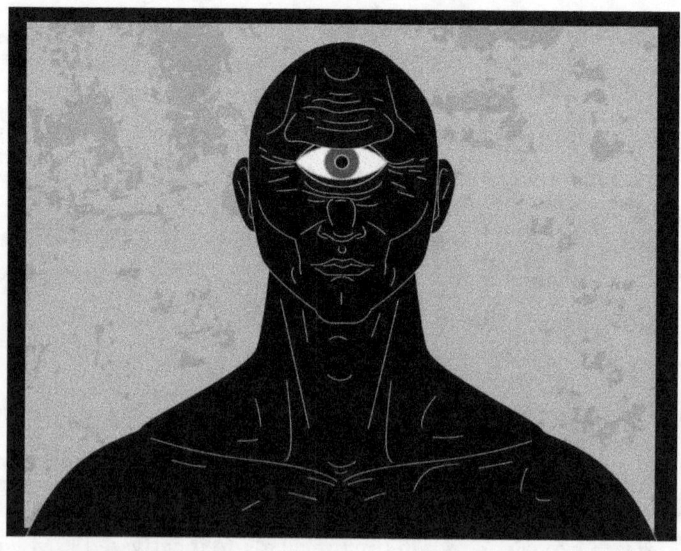

**Illustration 11:** The Third Eye- Shutterstock_1813753721

# Thirteen

# Human Consciousness And The Afterlife

*There have been over 20,000 papers or so written on the subject of consciousness and no consensus. Never in the history of science have so many people devoted so much time and produced so little.*

**Michio Kaku (1947 -) Japanese American theoretical physicist**

The reference to consciousness is relevant to Afterlife because human consciousness is identified with the mind and life.

If there is an Afterlife it follows that the mind and life of a person survives physical death.

In this context a distinction needs to be made between body, soul, and spirit.

The body and its working is within the realm of science. The soul is linked to the belief systems of religions e.g. Judaism, Christianity, and Islam believe in the existence of the soul body. The souls of humans, past and present will be resurrected on Judgement Day.

The spirit body concept relates to the belief systems of esoteric religions like Hinduism where on physical death the departing spirit reincarnates into a new body that chalks out a fresh role for itself based on past samsara and karma.

The process is believed to be endless till the attainment of moksha.

Some religious scholars are of the view that resurrection and reincarnation are two sides of the same coin!

Nobody can prove that resurrection and/or reincarnation will not happen.

However, the science purist would take the stand that we are sure only of the body, because its working can be scientifically analyzed and understood.

The renowned British philosopher, Bertrand Russell had this to say on the subject of afterlife:

"The water in a river is constantly changing though it flows in a channel on the riverbed. In a like manner, previous events have worn a channel in the brain, and our thoughts flow along this channel. This is the cause of memory and mental habits. But the brain as a structure, is dissolved at death, and memory therefore may be expected to be also dissolved. There is no more reason to think otherwise than to expect a river to persist in its old course after an earthquake has raised a mountain where a valley used to be."

There is a mathematical paradox that supports Russell's logic.

Global demographic statistics reveal 100 births and 255 deaths occur in a minute I.e.every minute there would be 155 babies in excess of the number of available souls!

Does this mean that some bodies can exist without a soul?

Nobody knows for sure whether we will be born again or whether we existed before birth.

Across the world's major religions there is a wide spectrum of belief on this arcane topic. The common factor is that none of the beliefs hold up to strict scientific scrutiny.

Between religions there are significant differences. As already stated the monotheistic religions, Judaism, Christianity and Islam subscribe to a Personal God and the fate of any individual past or present is decided in a once off Judgement Day. Those who pass the test are promised Everlasting Life and the failed ones Everlasting Damnation. The basis for evaluating the individual's conduct during his lifetime tends to vary with the religion.

In the religions of Asia, despite the plethora of Gods, Goddesses, God Men and God Women and their enormous number of adherents, the basic tenets are philosophical and the God concept is closer to the God of Spinoza popular with scientists the world over!

Unlike Hinduism, Buddhism believes in endless humanity and not endless individuals.Salvation for humanity as a whole depends on the attainment of *Nirvana* (total release from mental and physical suffering) through the practice of detachment and asceticism. Humanity is made up of individuals with repeating cycle but there is no continuation of individual identity through reincarnation!

Confucianism and Taoism are ancient philosophies steeped in the wisdom of the East that define a different way of life.

Confucianism accords little importance to the Afterlife:

The most important thing in his (Confucius) teaching is the relationship between humans and how people should behave in order to make the society better for people. He (Confucius) cares nothing about afterlife, gods, beginning of universe and things that never impact people's life.

Taoism follows a similar track:

DONT BELIEVE ANYTHING THAT TELLS YOU, GIVE UP THIS ONE LIFE YOU HAVE AND FOCUS ON AFTERLIFE OR NEXT LIFE, SIMPLY BECAUSE YOU HAVE NO IDEA WHERE AND WHAT YOU WILL END UP IN.

BELIEVING IN THAT OR SAVING FOR THAT (AFTERLIFE) IS REALLY NON SENSE.

SOME OF FALSE RELIGION ARE TRYING TO SELL A BLIND VIEW OF LIFE TO PEOPLE IN ORDER TO MAKE SOME PRESENT BENEFIT FOR THEMSELVES AND THAT IS NOT RIGHT THING TO DO.

THIS LIFE IS IMPORTANT AND TRYING TO MAKE IT GOOD HAPPY, ENJOYABLE,EVEN DO SOMETHING GREAT WITH IT WILL RAISE YOUR FUTURE ONES TO A DIFFERENT LEVEL.

Most of us embrace the religion we are born into and follow its rituals and practices throughout our lives without question.

# Fourteen

# Determinism And Free Will

*Man is a masterpiece of creation if for no other reason than that, all the weight of evidence for determinism notwithstanding, he believes he has free will.*

**Georg Christoph Lichtenberg (1742 – 1799) German scientist**

The significance of free will can be better understood by probing a few related words:

*Naturalism, fatalism, determinism, consciousness, awareness, predestination, panpsychism,* and *psychodynamics.*

We shall define these words, highlighting their relationship with free will:

Before we do so, for a better overall perspective, what is a layperson's understanding of free will?

With reference to humans, it is the ability *in a given situation to act freely in accordance with one's will or more specifically one's desire.* In the course of this narrative, we shall try to assess how far this definition of free will is realized in practice.

**Naturalism:** started as a literary movement spearheaded by writers like Emile Zola (1840-1902), Henry David Thoreau (1817-1862), and Thomas Hardy (1840-1928); their thinking was inspired by Charles Darwin (1809-1882).

It is based on a belief system that everything in the universe is a *single entity* driven by a *single cause.* This includes humans as causal objects whose lives, like the earth's orbit round the sun is predetermined to play out in a particular way. Naturalism does not support free will.

**Fatalism:** is almost synonymous with naturalism; closely linked with the Hindu belief in *samskara, karma,* and *reincarnation.* Humans are governed by the naturalism principle, but considered to fall outside the universal single entity matrix. This will be explained later in the narrative.

**Determinism and Predeterminism:** has a meaning very close to naturalism I.e.everything is *predetermined.* Determinism is the antithesis of free will.

It is to be noted that the supporting logic behind naturalism,fatalism, and determinism is slightly different.

**Consciousness:** We are familiar with words like 'fainting', 'swooning' in day to day life which are associated with becoming *unconscious.* These are abnormal medical conditions caused by factors such as *hypotension or syncope* (drop in blood pressure) or *hypoglycemia* (drop in blood sugar)

In this essay, we are focusing on the opposite positive state of *human consciousness* with particular reference to its role in determining the outcome of an event I.e. the *effect of the observer on the observed.* It does seem as per this proven phenomenon in physics that somehow the consciousness (free will) of the observer affects the nature of the observed; this suggests that free will in humans is a reality that cannot be ignored.

**Awareness:** is a *consequence* of consciousness, the awakening of our sense organs: Ocular (sight), auditory (hearing), olfactory (smell), integumentary (feeling), and gustatory (taste).

With the sense organs, humans are able to perceive events in external world within certain limits, interpret and act on them with the brain as coordinator. This sequence is indicative of the existence of free will.

**Predestination:** is the antonym of free will. In the Hindu school of thought however, predestination and free will can coexist; faced with a problem with multiple solution options, though predestination predominates,the individual can with spiritual help reduce its effect. This can reflect in his final decision and action in certain situations. In such cases, he is responsible for the decision he makes and the resultant action.

**Panpsychism:** states that all things have a mind like quality. The question that logically follows is, what are *all things* and *mind like* quality?

'All things' include all animate and inanimate objects in the universe. Defining 'mind like' quality is akin to exploring an abstract concept open to many interpretations.

A popular theory links panpsychism with consciousness (free will) which is difficult to accept in the case of inanimate objects and lower biological life forms like plants.

Consciousness requires the functioning of a brain, ideally of the human type. As we go down the biological ladder it is almost impossible to identify a cut off point for consciousness.

It is however fair to say that panpsychism is applicable to humans.

**Psychodynamics:** relates to the relationship between *conscious motivation* and *unconscious motivation*.

Another definition relates to a connection with Sigmund Freud (1856-1939) psychoanalysis theory based on psychological energy (libido). The reference to conscious motivation and/or psychological energy is suggestive of support for free will.

The above definitions of words related to free will, indicate that they are based on closely related belief systems. Even so there is no clear unanimity whether free will is present or absent in biological life forms such as the human life form.

We shall turn to the spiritual world in the hope that it can provide convincing answers.

1.  **Judaism:** Talmud is categorical regarding the presence of free will in humans:

    *How precious is man, created in the image of God* (Talmud-Avot3:18)

    The belief is that since God has free will, *ipso facto* man too has free will; this is part of God's master plan to shape and change the world through man His supreme creation. Talmud further clarifies:

    *Greater than the gift of free will, is that God **told us** we have free will.*

(Talmud-Pirkei Avot 3:19)

It is made clear that the gift of free will that God confers on man is not a free for all license, to indulge in material pleasures and merry making. Man needs to distinguish between cravings of his body and the higher aspirations of his soul. The latter is God's will, with which he needs to identify, work towards and fulfill, thereby making proper use of free will.

2. **Christianity:** The bible deals with free will in two ways:

a) Theology: It is stated that all who wish to be redeemed and exercise their free will accordingly, are not necessarily accepted. Consider the following verses appearing in succession, on statements made by Jesus:

*No one can come to me unless the Father who sent me draws him. And I will raise him up on the last day* (John 6:44) and

*It is written in the Prophets, 'And they will all be taught by God.' Everyone who has heard and learned from the Father comes to me* (John 6:45)

The first verse is restrictive in respect of an individual qualifying for redemption, whereas in the very next verse it is stated that there are no conditions and redemption is open to all. Can this apparent contradiction be reconciled?

Biblical scholars have clarified that the Greek translation of the word *draw* used in the first verse is *impel* or *inspire* which means that man cannot qualify for redemption by his free will alone, but needs to be impelled or inspired by God (Father) to seek this option.Does this mean that despite exercising the free will option for redemption there would be the possibility of being rejected?

Biblical scholars have turned down this possibility.

Ephesians I:4-5 says:

*God chose us in him before the foundation of the world, that we should be holy and blameless before him. In love he predestined us for adoption as sons through Jesus Christ according to the purpose of his will.*

This refers to a timeline before the world was even created!

In summary, we conclude that the Christian belief is that *all those* who as per their free will seek to be redeemed from sin through Jesus Christ will be accepted on the day of reckoning in the kingdom of heaven. There are no exceptions to this rule.

b)   Conventional: This is the commonly understood usage I.e. the application of one's free will on mundane problems that arise in day to day living. This is a reality for all of us, irrespective of the religion we follow.

We exercise our free will and make judgements and decisions when confronted with multiple options. We assume responsibility for all our actions whether financial, familial, health related, unexpected reverses, deception, let downs. During a lifetime, such events pursue us relentlessly, and in each case we find a way out. Even favourable turn of events like a winning lottery ticket require choosing between different options!

The Christian position is the same as in Judaism, that *God created humans in his image* (Genesis 1:26). Unlike animals who respond to instinct, humans are endowed with love and the spirit of justice. His superior brain enables him to make free choices to uphold these values. Success or failure depends on us and not on fate.

*All that your hand finds to do, do with your very power* (Ecclesiastes9:10) also

*The plans of the diligent one surely make for advantage* (Proverbs21:5)

Again in the New Testament of the bible:

*Free will is a precious gift from God, for it lets us love him with our **whole heart** because we want to* (Matthew22:37)

To summarize, free will is a gift to man from God to be used for the good of all.

3) **Islam:** There are several verses in the Koran which support both free will and fatalism. I have chosen three of the most striking from the two categories:

Free Will:

a) *Nay, it is surely a Reminder. So whoever pleases may mind it* (Koran: 74.54/55)

b) *The truth is from your Lord: let him then who will, believe; and let him who will, be an unbeliever* (Koran: 18.28)

c) *We have truly shown him the way; he may be thankful or unthankful* (Koran: 76.3)

Predestination (Fatalism):

*Taqdir* or the absolute decree of good and evil is the sixth article of the Islamic creed, an indication of the Divine Will which is recorded on the preserved tablet and believed to be final and irrevocable. More specifically:

a) *The Lord has created and balanced all things and has fixed their destinies and guided them* (Koran: 87.2)

b) *By no means can anything befall us but what God has destined for us* (Koran: 9.51)

c) *No disaster occurs on earth or accident in yourselves which was not already recorded in the Book before we created them* (Koran: 57.22)

Summarizing, despite the verses from the Koran quoted under Free Will, it is predestination that prevails.

This viewpoint is further reinforced towards the end of Muhammad's days I.e. of uncompromising fatalism.

4. **Hinduism:** To understand the Hindu viewpoint, we need to unravel the meaning of key Sanskrit words: *prarabhda,purushartha,ashrama vyavastha, daiva,*and *purushakara.*

   a) Prarabhda karma relate to past actions that influence our present life and future incarnations. It is essentially pay back time I.e. whatever you 'sowed' you will now 'reap'

      Prarabhda karma literally means *action that has begun* (retribution).

   b) Purushartha refers to the four stages of the human life cycle at a *moral* level: the first is 'dharma' (good thoughts and deeds), and the last 'moksha' (Hindu concept of salvation); in between are 'artha' (money) and 'kama' (worldly pleasures).

      To attain moksha, the hurdles of artha and kama have to be crossed.

   c) Ashrama Vyavastha, also refers to four life stages but at a *physical* level: 'Brahmacharya'(student), 'Grihastha' (householder), 'Vanaprastha' (retired) and 'Sanyasa' (renunciation)

      As per Hindu belief Purushartha and Ashrama Vyavastha are closely linked.

   d) Daiva and purushakara can be taken together: daiva is luck or fate, purushakara is human effort.

Karma, fate and free will and their conflicting claims when related to individual humans have baffled religious scholars and philosophers for centuries. The ancient sages of India understood the underlying reasons for the dichotomy and were able to see through it. They were able to take a middle path and reveal how fate ('daiva')and free will ('purushakara') could coexist.

Astrology, as a guide to an individual's future is a factor which favours 'daiva' since the influence of the planets and moon phases on an individual's life is clearly beyond human control. The reliability of astrology as a science or pseudo science has been researched and has many followers even today, in India and a dozen countries across the globe; the list headed by India include United States,Canada,United Kingdom,Australia,South Africa and Singapore.

In addition there are many websites and online practitioners with adherents fro diverse countries.

Apart from planetary influences there is the belief in 'karma' 'samskara' and reincarnation which individually and jointly believe in future lives and the inevitability of retribution applying to an individual's present and future lives till 'moksha' is achieved.

With such strong arguments favouring 'daiva' that arise out of an individual's 'prarabhda' does Hinduism have any place for free will?

The answer surprisingly is *yes*. Reasoning behind this viewpoint is:

Humans constitute the zenith of God's creation and as such are endowed with sufficient freedom to make or mar their future. The spark of divinity gives an individual space to counter the effects of 'daiva' and 'prarabhda' and partially mitigate them. Summarizing, Hinduism does not support *merciless fate* but rather a blend of *free will and fatalism* in all cases based on morality.

The brief summary of belief systems related to free will in humans and the viewpoints of major world religions serves as an eye opener.

We tend to take *free will* in humans literally, at its face value and fail to see its deep hidden meaning and significance.

Let me end with the following quote from the German philosopher Arthur Schopenhauer (1788-1860):

*Man can do what he wills but he cannot will what he wills!*

Humans are unable to reach a consensus on free will. The mystery persists.

# Fifteen

# Proprioception

"Thought cannot be separated from me, therefore I exist". "COGITO TERGO SUM" translates to

"I think, therefore I am"

THE FIRST PRINCIPLE OF DESCARTES

Rene Descartes French philosopher (1596 - 1650)

I have commenced the essay with a profound quote from Rene Descartes of the 16th century. It is appropriate to do since the topic "Proprioception" though focussing on body position and movements is intimately connected with the mind. It also relates to the innumerable sensory nerve endings in the body from crown to toe.

God as a sculptor, created the ultimate when he made "man" with special abilities way above other life forms. Proprioception is one such special and subtle ability.

A pertinent digression on the relevance of prosthetics:

HAPTIX an acronym for "Hand Proprioception and Touch Interfaces" is a US government funded research programme with emphasis on "prosthetics". It attempts to bridge the gap between movement and sensation in prosthetic arms. Using external sensors to sense pressure, movement, or texture, HAPTIX aims to create a safe and reliable interface by connecting sensors to the nerves that are left above the point of amputation. In a recent study, researchers implanted tiny cuff electrodes around the three main nerves of the forearm. Cuff electrodes that wrap around the outside of a nerve is the preferred option to electrodes that pierce the nerve and possibly cause damage.

Through multiple channels on each electrode, electrical impulses were applied with varying intensity and frequency.

In the lab, implanted amputees were wired to a computer so that the researchers could experiment with different stimulation patterns. Thus, they were able to create sensations like tingling, pressure and texture.

It was found that amputees showed dramatic improvement with delicate tasks when they were wired in to feedback from sensors on the prosthetic finger.

HAPTIX has still a long way to go. Wireless electrodes will allow for an "untethered" prosthesis. Refined nerve stimulation is needed to achieve precise sensations required for proprioception. Sensitive prosthetics offer benefits beyond improved control. For instance, phantom pain is a common complaint of amputees: painful sensations that seem to come from the missing limb. Strangely, phantom pains seem to subside if you give the sensory nerves something to do. Proprioception and touch should also restore "embodiment" helping amputees to view their prosthetic as a true extension of their body than just a fancy tool. This sheds light on an important role of proprioception. It seems that sensation and muscle sense in particular is crucial for a healthy metaphysical self image. In the experience of one patient, a senseless arm is viewed as a completely alien entity. In anonymity case bedtime was distressing because closing his eyes left the patient feeling disembodied and "floating"

It seems that Descartes principle could be rephrased as "I sense therefore I am."

Proprioception (PP in short) is sometimes referred to as 'sixth sense' but with a connotation different from 'sixth sense' as used in common parlance!

Usually, the sixth sense is understood as a paranormal power in living creatures, man in particular, that transcends the five senses of sight, hearing, touch, taste, and smell that enable the normal awareness of the the world we live in.

For example phenomena like ESP (Extra Sensory Perception) fall in the category of 'sixth sense'

Proprioception is *not* ESP.

It relates to the imperceptible role played by the motor and sensory nerves of the human body in giving it a sense of depth and balance.

Kinesthesia?

Kinesthesia is a related phenomenon but distinct from PP. It pertains to the awareness of movement (up, down, in, out) and range while a joint in the body is in motion like the swinging arm of a bowler in cricket (initial, mid, terminal).

An impaired kinesthesia indicates dysfunction in the peripheral nerves, spinal cord, brainstem, or cerebrum.

Proprioception on the contrary, is joint position sense and awareness of joints "at rest". During a test for PP the examiner moves the subject's extremity/joint through a range of motion and then holds it in a static position. The subject is asked to pinpoint the position. Inability to do so indicates that the PP is abnormal and the individual suffers from "sensory ataxia" (Impaired sensation due to nerve damage. This causes less feedback from your brain telling you where your body is in relation to the ground)

The instances of kinesthesia and PP described above though interesting are not self explanatory. We shall probe a few commonplace events that better exemplify the phenomenon of PP:

Actually, proprioceptors are tiny little nerve endings in our bodies that help with balance, performance and setting muscle tension. Very few people unless seriously injured, like loss of limb would have heard of PP. However it is with us all the time, day in and day out.

When we walk, we take for granted all the little things that must take place to get from point A to point B without falling. The eyes play a major role along with our "vestibular system" (explained later) in maintaining body balance. What most people don't realize is that we have little nerve endings in our hands and feet that tell our muscles and the rest of the body exactly where we are in position, space and time. This means you can close your eyes and touch your nose with your finger tip because of proprioceptors. Those little nerve endings know where your arm is in space allowing you to perform the desired action.

Now, the real test- Try standing on one foot and get your balance. Not too hard for many, but impossible for some. Now close your eyes while standing on one foot. Once the eyes are closed, you might feel your foot shaking from side to side. That shaking is coming from the little nerve endings firing away to maintain balance.

Whenever we step, the nerve endings are performing just like when your eyes were closed, but we fail to notice this because of our visual and vestibular input. We don't fall over because those nerve endings recognize that we are on solid ground and our muscles adjust appropriately.

Why is this relevant?

The elderly, particularly those with an injury have less reactive proprioceptors. More often than not, when we hear of someone older falling it wasn't because they got light headed, but rather due to the proprioceptors failing to communicate properly. If they fail to fire when they are supposed to, the body can't tell the difference between flat ground and uneven ground. More injuries occur this way because the body reacted like it stepped in a pothole or the floor suddenly moved beneath unstable feet.

Proprioceptors act like the balance poles used when walking a tightrope. It helps maintain a stable centre of gravity. If we are falling to the right, the pole balances the left and vise versa.

Poor proprioceptive activity leaves the body walking a tightrope without a pole. Unfortunately, for many people there aren't safety nets to catch us when we fall. This leads to fractured bones and other injuries.

The young aren't exempt either. An injured area loses the acuity of proprioceptors. A common example is a sprained ankle. This happens because proprioceptors failed to provide stability showing even ground as dangerous unstable environment. With each step, certain muscle fibres have to relax and contract due to the surface beneath it. This happens automatically, unless we have flaws in our neuromuscular system. All it takes is slight misfiring and we can lose balance or have a joint suddenly give way making us prone to injury.

Now for the good news:

Proprioceptors can be retrained at any age to regain control. Specific balance training and exercise can improve proprioceptors in a very short time. Athletes have reported a distinct improvement in performance following PP training.

A simple proprioceptive evaluation can easily detect if a patient could benefit from PP training. This should be undertaken in good time. As in other medical care situations, prevention is better than treatment and cure.

## THE CIRCUS:

Transfer your thoughts to the circus; more specifically the athletic men and women performing the "trapeze" act. What enables the incredible confidence and precision in the artists that leave the spectators spellbound and breathless?

These are the words of a spectator in a circus in Paris:

"The acts that put me in awe were the rope and aerial dancing. Their coordination and ability to move so smoothly is fascinating. Also the tight rope walker. who not only did stunts on the ropes but managed to walk across in high heels. I have tried them and find myself tripping over my own feet on solid ground, let alone on a 1 inch rope. It is quite amazing how these performers have perfected each of their moves with such ease, even when flying through the air or walking on a rope."

## SKYDIVING AND HANG GLIDING:

The significance of this sport goes way beyond the high fives of a happy landing. It is your proprioceptive sense, a neuromuscular phenomenon that allows you to know where all the parts of your body are in space. It gives you the ability to safely maneuver your body around your environment. It gets your parts working as a whole, strengthens those parts in the most integrated way and keeps your whole from bumping into everything in the immediate vicinity. PP in short is "grace".

Skydiving is particularly good in developing PP in the trainee performer.

It's super power lies in the fact that the body is removed from the standard "feet on the ground" orientation and held there by the airflow, with the air molecules exerting force on the limbs and directing the position of the body. It's a workout to move the body around up there, for which excellent physical awareness is necessary to move through the progression of the skydive.

Skydiving in and of itself is not a treadmill running kind of workout. However, carrying heavy gear around, packing your parachute, pacing through jumps with your buddies, running back and forth to fetch bits and bobs from the sidelines, striding back from your landings... It adds up.

More importantly, a skydive sharpens meditative focus on PP, i.e. our ability to have a sense of our own body parts in relation to one another and in space. It is essentially how our body sees itself and the world.

How exactly does this work? To understand the phenomenon better, we need to define a few technical terms:

"Muscle spindles" are proprioceptors that consist of "intrafusal muscle fibres" enclosed in a sheath (spindle). They run parallel to the "extrafusal muscle fibres" and act as receptors that provide info on muscle length and rate of change in muscle length. The spindles are stretched when the muscle lengthens.

"Fusal muscle fibres" Extrafusal muscle fibres comprise the bulk of muscle and form the major force generating structure. Intrafusal muscle fibres are buried in the muscle and contain receptors for stretch, but they also contain contractile elements.

"Golgi tendon organ" is a proprioceptor, sense organ that receives info from the tendon that senses TENSION. When you lift weights, the golgi tendon organ is the sense organ that tells you how much tension the muscle is exerting.

"Vestibular system" is one of the sensory systems that provides your brain with info about balance, motion, and the location of your head and body in relation to your surroundings. It consists of two structures of the bony labyrinth of the inner ear, the vestibule and the semicircular canals, and the structures of the membranous labyrinth contained within them.

"Patellar Reflex Test" also known as the "Deep Tendon Reflex Test" is used for checking normal neurological reaction in a patient. The neurologist observes and interprets the reaction of a patient to a tap on his patella (knee cap) with a rubber hammer. A diminished reflex is associated with a peripheral nervous system (PNS) disorder while hyper reflexive responses are related to central nervous system (CNS) disorders. To develop PP successfully the PRT needs to be normal.

When a particularly situation requires PP to fire, information on joint angle, muscle length, and muscle tension are conveyed to our brains that integrates it with our "vestibular" system. This helps us in balance and spatial orientation and prevents possible injury. We shall probe the PRT or "patellar reflex test" further. The familiar tap on the knee causes a stretch in the "intrafusal" fibres that mimic the "extrafusal" fibres and relay that info to the spinal cord. An "alpha motor neuron" located in the spinal cord will then conduct an electrical impulse back to the quadriceps to contract the muscle, which is what leads to the kicking motion of your leg. This whole system exists throughout our body to prevent injury to our muscles by excessive stretching or contraction.

## WORKING OF PROPRIOCEPTION:

The following 'thought experiments' are revealing:

1.  You are sitting alone on a swivel chair in a room that is pitch dark. You swivel the chair which stops after a few revolutions.

    A voice commands:

    'Touch the tip of your nose with your right hand middle finger. Repeat the action immediately with your left hand middle finger'

    How many of us will get this right in the first attempt? Maybe 10%?

    If the room is lighted probably 90% will get it right in the first attempt

    The explanation lies in PP.

    The motor /sensory nerve communication highway between the hand and brain (command and control centre) is cut off in the dark room (compromised vision) giving poor PP and high failure rate.

    PP is restored in the lighted room giving a high success rate!

2.  Assume you meet with a serious road accident in which you lose half your right arm and half your right leg. Your right arm and right leg are reduced to disfigured stumps of flesh!

Following months in the orthopedic ward of a major hospital you return to your home with a modern Jaipur arm and leg that are not even visible to an outsider when the patient dons trousers and a full sleeve shirt.

He has also been trained in the hospital on the use of prosthetic limbs. A voice commands:

'Pick up the glass of water on the table QUICKLY' with your right hand'

If you literally attempt this simple action with a prosthetic limb you will almost certainly hook the glass of water off the table onto the floor.

Why?

The motor/sensory communication between the prosthetic hand and brain is poor because of the urgency of the action!

The PP improves if the patient is allowed to concentrate; he will then perform better!

A similar experiment can be done with the prosthetic leg with identical results.

For this reason many patients with prosthetic limbs get frustrated over time. They restrict the use of prosthetics and in some cases give them up altogether!

The muscles in the human body from crown to toe are embedded in nerve cells and an efficient PP (efficient communication highway) between any part of the body and the brain enables healthy humans, sportspersons and athletes in particular, to perform incredible physical feats.

The importance of an efficient PP cannot be overstated!!

# Sixteen

# Languages The Tower Of Babel

*Language is a process of free creation; it's laws and principles are fixed, but the manner in which the principles of generation are used is free and infinitely varied. Even the interpretation and use of words involves a process of free creation.*

**Noam Chomsky (1928 - ) American Linguist**

The origin of language is an eternal puzzle; unlike all other living beings, humans alone have the remarkable capability of conveying the most complex ideas and emotions to each other through the medium of language.

Dogs *howl*, hyenas *laugh* cats *purr* and horses *neigh* but none of them can 'speak' to each other in a 'language' based on an extensive vocabulary, idiom, grammar and syntax!

Is language a gift to humans from God or is it another instance of human superiority over other living beings as a fall out of evolution?

NKJV Genesis 1:27 of the Old Testament of the bible says:

*So God created man in his **own** image; in the image of God He created him; male and female He created them.*

Scholars have debated the likely meaning of the above verse:

Should it be taken literally? If so the interpretation is that it refers to the substance and form of Man vis a vis God. This interpretation is reinforced by the following verse:

KJV Genesis 1:28 that says:

*And God blessed them and God said unto them, "Be fruitful and multiply, and replenish the earth, and subdue it; and have dominion over the fish of the sea, and over the fowl of the air, and over every living thing that moveth upon the earth"*

There seems to be little doubt that God in His wisdom had created Man as a replica of Himself for a leadership role over other living beings on earth.

How does language fit into this framework?

It seems inconceivable that God would create Man for a leadership role without gifting him the power of language!

In the Christian Era, Jesus Christ, who lived for a little over 30 years in human form on earth, acknowledged by Christians as the Son of God says:

*I and the Father are one* (KJV John 10:30) an indication that the word *image* should be taken as per it's literal meaning.

From Adam and Eve through the 5-6 millennia of the Old Testament and into the Christian Era there are innumerable instances that indicate human communication through the medium of language!

For example in the Garden of Eden God had specifically communicated to Adam that they could partake of any fruit except that borne by the Tree of Knowledge of Good and Evil; subsequently Adam was tempted by Eve through the machinations of Satan to violate this instruction.

Such a sequence of events could not have taken place without the use of a language!

This view is supported by experts like American linguist Noam Chomsky and neuroscientist David Poeppel that language is hardwired in the human brain since inception and is not a product of evolution.

There is an alternative explanation:

Since God is *omnipresent* I.e. present in everything, it is possible the word image is referring to *nature* I.e. man is created like all other living forms but enjoying an exalted leadership status.

Such an interpretation is more in line with the Koran that Allah (PBUH) is unique and cannot be likened to anything in the universe including man.

There are members of the scientific community who favour the latter interpretation since the concept of *Personal God* is replaced by the *God of Spinoza* (more aligned to *Creation* than a *Creator*)

Be that as it may, there is little doubt that God and the earliest humans Adam and Eve did communicate with each other, and from the nature of the subject matter and it's nuances, it seems highly unlikely that this took place through means other than the medium of a language!

## There is a related quandary:

There are today an estimated 6000 languages in the world that can be divided into about 20 language families. What could be the explanation for the *multiplicity* of languages assuming that language is God's exclusive gift to humans.

## THE TOWER OF BABEL:

The exhortation by God to humans 'to be fruitful and multiply'......... (Genesis 1:28) is repeated to the survivors of the Flood in Noah's Ark. The rebellious leader Nimrod, with the help of Satan defied the edict, shut himself and his followers in a tower. This was an easy and safe life as compared to the nomadic conquering lifestyle implied in the edict.

Genesis 11:6 says:

*Nimrod, with direct access to demonic intelligence and Satanic power, would be invincible without divine intervention.....*

*In God's judgement, the main problem was the unity of the people; the one most effective way of thwarting unity would be to prevent communication!*

The people gathered in the tower began to speak in different tongues. Even today when there is confusion arising out of multiple languages, it is referred to as a 'Tower of Babel'

Use of multiple languages within the tower ensured that the Flood survivors as a group became dysfunctional since they could not understand each other. Sub groups separated as per their language and dispersed far and wide where individual languages gradually expanded their vocabulary, developed a script, rules of grammar and syntax!

## ONOMATOPOEIA:

Some of us would be familiar with this *figure of speech*.

It is best explained by citing a few examples:

*crack, growl, gurgle, roar, whoosh,* and many others. The common factor in these words is that the *sound of the word is indicative of it's meaning.*

A detailed study of onomatopoeia words reveals that they can be divided into 5 categories:

1.  Air Sounds: *gasp, wheeze, fizz, thud, squish.*

2.  Water Sounds: *splash, spray, drizzle, squirt, whoosh*

3.  Collision Sounds: *ticktock, crack, screech, slap, clap*

4.  Vocal Sounds: *mumble, murmur, bawl, giggle, grunt*

5.  Animal Sounds: *hiss, meow, bray, moo, buzz*

It could be argued that early man, like the Neanderthal, who lived 50,000 years back coined words like the above from the way they sounded and assigned meanings to them accordingly. The birth of words and a vocabulary grew in this manner. The initial use of words arose out the sheer need to survive through communication through a medium that came to be known as a *language.*

With the development of 'homo sapiens' over several millennia in respect of body characteristics and brain, 'languages' that originated as onomatopoeia words expanded their vocabulary, developed scripts, rules of grammar, idiom and syntax leading to the highly refined languages of today.

Human migrations over vast expanses with different climatic conditions, environment, and terrain resulted in multiple languages developing simultaneously.

I have briefly summarized the thinking of the evolutionist that explains the plethora of languages that exist today and the high level of refinement in many of them.

God given or product of evolution?

WE DO NOT KNOW!

We shall probe some of the world's ancient languages for clues regarding the *origin* of languages. The oldest languages in serial order are:

SANSKRIT

CHINESE

GREEK AND LATIN

HEBREW

The basis for shortlisting of these languages is entirely subjective. There could be alternative lists as a basis for meaningful case studies:

e.g. EGYPTIAN (ARABIC), ARMENIAN, EURO ARAMAIC, and TAMIL is a respectable alternative list!

We however intend to stick to the first list for the probe:

SANSKRIT: is often referred to as the *mother of all languages!*

Mother in a general sense is the originator of all life forms; in the case of humans it includes the phenomenon of languages.

A study of the origin of the 'mother of languages ' I.e. Sanskrit would logically be a good place to gather clues on the origin of *all* languages.

## ORIGIN-THE CREATIONIST VIEWPOINT:

The discussion that follows, seeks evidence from the world's oldest languages for the *creationist* viewpoint of the origin of languages I.e. as a gift of God:

But does Sanskrit the first *spoken* language qualify as the mother of all languages? (I have used the word *spoken* because common sense tells us that the script and written language would have emerged only much later)

Historical records reveal that there can be no universal consensus on nominating the 'mother of all languages'

There would be few takers for the theory that the first humans, the biblical Adam and Eve spoke to each other in Sanskrit!

But were the biblical Adam and Eve going back 5-6 millennia the first man and woman? Judaism, Christianity and Islam would support this theory; however Asian religions like Hinduism in which the ancient *Vedas* (conversations between *guru* -teacher and *shishya* -pupil) of Hinduism 4-5 millennia back support the probable use of Sanskrit!

A test that could be applied to ascertain the age of a language is the range of its vocabulary I.e. larger the number of words, older the language.

## VOCABULARY:

Sanskrit is estimated to have 100 billion words! This is an astonishingly large number and seems unbeatable; however Mandarin Chinese *character* combinations, recognized as distinct words could theoretically yield a *countless* number!

The Hebrew language has less than 100,000 words. Yet along with Greek it is generally accepted as the language of the the Old and New Testament of the Hebraic bible that go back 5-6 millennia in history

(To put this in perspective the English language which does not make any claims of being the world's oldest language has in theory about a million words. However most of them are rarely used. Shakespeare created his masterpieces in just 20,000-30,000 words)

The number of words line of reasoning is clearly endless and not leading to any definite conclusion.

In this context following words of noted British philologist William Jones (1746-1794) in 1788 are relevant:

*The Sanskrit language, whatever be its antiquity is of a wonderful structure; more perfect than the Greek, more copious than the Latin, and more exquisitely refined than either, yet bearing to both of them a strong affinity a stronger affinity, both in the roots of verbs and the forms of grammar, than could possibly have been produced by accident; so strong indeed, that no philologer could examine them all three, without believing them to have sprung from some common source, which perhaps, no longer exists.'*

This suggests the existence of a *prehistoric ancestor language* of the Indo European family of languages that is now extinct.

Be that as it may, we shall continue our narrative on the basis that Sanskrit is the 'mother of all languages' though only arguably so.

## SCRIPT:

The first evidence of written Sanskrit script emerged around 300 B.C.E. in the form of the *Brahmi* script used first in the *Prakrit language;* However, linguistic scholars consider Prakrit as a descendant of Sanskrit strongly suggestive that the Brahmi script had probably already been used earlier in Sanskrit! There is no recorded evidence to show who created the script and how it developed. Modern Sanskrit is written and taught in the *Devnagri* script used in written Hindi.

The case study of Sanskrit one of the oldest languages of the world, reveals that the origin (vocabulary and script) is shrouded in antiquity; perhaps as suggested by Jones, it had its origins in an ancient 'root language ' about which little is known and probably extinct'

A study of the Sanskrit language does not seem to lead to any firm conclusion on the likely origin of *all languages*.

As a gift of God to humans is more an article of faith with Creationists than a research finding that can hold up to scientific scrutiny!

## CHINESE:

ORIGIN:

We go back to 2500 B.C.E i.e. 5000 years into ancient Chinese folklore for clues.

The Chinese believe that their written and spoken language occurred *simultaneously* and is of *divine* origin. The originator is Godlike and referred to as *CANGJIE* and is believed to have had 4 eyes!

This supports the Creationist view that the gift of language emanates from divine sources.

This is at variance with present day China where majority of the population while respecting the philosophical teachings of their forbears are either atheists or at best agnostics.

(In this context in neighbouring India, the mythical *SHIVA* the God of Destruction in the Hindu trilogy is depicted to have had 3 eyes, the normal two and the third in the middle of the forehead that lies closed and dormant most of the time)

## VOCABULARY:

Tracking the vocabulary of the Chinese language is an extremely difficult task:

In the first place, when we mention the *Chinese language* in common parlance, we are unknowingly stating a classical euphemism!

Which Chinese language?

As stated by Jawaharlal Nehru in his book *Discovery of India*:

"Ancient India like *ancient China was a world unto itself, a culture and a civilization which gave shape to all things*"

In terms of size, China is the third largest country in the world trailing Russia and Canada. Within the borders of the China of today, there are gigantic mountain ranges (the *Himalayas, Karakoram, Pamir Mountains*) mighty rivers (*The Yangtze, Yellow River, the Heilong Jiang*) and huge deserts (*Gobi Desert*)

Such geographic features resulted in the creation of natural regional borders where the earliest nomads from neighbouring Mongolia, the Xanbei tribe,the Manchus had settled and formed distinct communities with unique dialects over large swathes of time.

Thus arose the significant variations in the ethnicity of the Chinese people, their customs and languages!

For starters, with reference to languages, there are the two broad categories of *Standard Mandarin* (official language of Mainland China and Taiwan) and *Cantonese* (Hong Kong)

Standard Mandarin is also the official language of Singapore and one of the six official languages of the United Nations. In spoken and written form it has 1.2 billion adherents, the highest in the world for any language.

The difference between spoken Mandarin and Cantonese is indeed very striking!

(From an Indian point of view the two stated categories are as different from one another as Punjabi spoken in the Northern part of India and Tamil in the South!)

You can have two Chinese in the same room, one speaking Mandarin and the other Cantonese and they would have no clue what the other is saying!

And yet they use the *same* written language- the *same script!*

## SCRIPT:

Here again there are surprises:

There are no alphabets in the Chinese script-instead there are *characters* (50,000) Most of these are not used. To put it in perspective, the numbers in use varies from 3000-8000 depending on the requirement of the user (3000 to read a newspaper extending to 8000 for writing a thesis for a PhD)

Each character which is a combination of six fundamental brush strokes could represent a syllable, a word, or a combination of words having both a meaning and possibly *a hidden meanings!*

Not surprisingly this gave rise to large pockets of people each with its own distinct vocabulary, grammar and syntax spawning innumerable *dialects*

The most important of these is the Standard Mandarin also referred to as the *Beijing Dialect.*

To complicate matters further, the innumerable *dialects* have different *tonal qualities* e.g. Mandarin has 4 that could go upto 10 in other dialects!

The tonal quality can mean different meanings or nuances for the same word in different dialects that create district nuances!

*A picture is worth a thousand words* is a popular Chinese saying:

The converse is equally true- each Chinese character is a picture or *pictograph*.

All this may sound perplexing to an English audience; however to a Chinese student trying to learn English, the English language is mind bogglingly confusing.

The development of the Chinese script that are presently used in the innumerable dialects throw little light on the fundamental question whether it is God given or the result of Evolution.

## GREEK AND LATIN:

ORIGIN:

In common parlance we often say:

*It is all Greek and Latin to me!*

We say this when faced with a topic that is difficult to understand or on hearing an unknown language!

Though in many cases unintended there is a distinct derogatory undertone in the statement.

The reason for this could be that despite their long history, Greek and Latin today are not in the limelight.

Greece is considered the *sick man* of Europe, with acute financial problems requiring subsidies from the European Union.

Latin has a rich ancestry but spoken sparingly and generally considered an *extinct language.*

Be that as it may, Greek and Latin are linked to two of the oldest civilizations on earth (Greek and Roman)

Landmarks in Greek civilization date back to the hoary days of Homer (author of the *Illiad and Odyssey* circa 800 B.C.E-700 B.C.E)

Scholars believe that the destruction of Troy and the Trojan war in Ancient Greek history happened even earlier circa 1200 B.C.E.

Greek was the lingua franca of the period and is believed to have existed even earlier in oral form.

The big three of ancient Greek philosophy- Socrates, Plato and Aristotle are estimated to have belonged to a later period i.e. 500 B.CE-300 B.C.E.

There were less famous Greek philosophers of an earlier period:

Thales of Miletus- circa 624 B.C.E-546 B.C.E

Pythagoras of Samos- circa 570 B.C.E- 495 B.C.E

Anaximenes- circa 550 B.C.E

Heraclitus- 500 circa B.C.E

This catalogue of highly evolved scholars and philosophers indicates that the Greek language of those times was highly refined.

Little is known of how the Greek spoken language, written script (including the formidable Greek alphabets and symbols used in present day mathematics) grammar and syntax *actually evolved.*

We do have some leads- Widely accepted versions of Greek in chronological order are:

Proto Greek (Early Hellenistic period circa 3200 B.C.E)

Mycenaean Greek (Greek mainland, Crete and Cyprus circa 1600 B.C.E-1200 B.C.E)

Ancient Greek (Ancient Greece circa 900 B.C.E-600 B.C.E)

Koine Greek (Alexander the Great circa 400 B.C.E)

Medieval Greek (Byzantine Empire circa 400 B.C.E)

and finally Modern Greek (Greece. Europe, Middle East of the present day)

A similar reconstruction of Latin reveals the history of the Latin spoken language and written script.

Unlike English (SVO *Subject Verb Object*) Ancient Greek and Latin followed (SOV *Subject Object Verb*) syntax. An example from English of the present day would help:

*Mujeeb likes mutton curry* (SVO in English, French, Mandarin, Spanish, Russian)

*Mujeeb mutton curry likes* (SOV in Sanskrit, Ancient Greek, Latin)

*Likes Mujeeb mutton curry* (VSO in Biblical Hebrew, Classical Arabic, Irish, Welsh)

Some languages permit use of both SVO and VSO syntax structures!

## VOCABULARY:

Greek and Latin are derived from the Indo European family of languages. Greek is the language of Greece and countries in and around the Mediterranean sea like the Greek mainland and Cyprus. The use of Latin is restricted,almost in total disuse. Vocabulary presently in use in countries in the Mediterranean region like Italian, Spanish, French and further west like English have many of their roots in either Greek or Latin.

For example 'Father' in English is derived from *Pater* (Greek and Latin) 'Brother ' from *Phrater* (Greek) and *Frater* (Latin)

The Romans used Latin as their medium of written and oral communication during the long tenure of the Roman Empire of 1500 years circa 30 B.C.E -1500 A.D. extending almost to the whole of present day East and West Europe.

## GREEK SCRIPT:

The Greeks were the first people who introduced the concept of *alphabet* i.e. a single letter form that denotes a sound with which words can be formed with a wide range of meanings. The words would serve as the *building blocks* of a language.

Earliest records of written Greek go back to 1500-1200 B.C.E when the Mycenaeans (ancient Greek tribe) attempted a script based on *Minoan* syllabary. This was not successful because the Greek vocabulary of the time was found to be incompatible with the script.

In 900 B.C.E Greeks used the *Phoenician* script as a basis for developing variants suitable for the different spoken dialects:

Examples: The modern Greek letter **KAPPA**(K) has remained unchanged in the *Ionia, Athens, Corinth, Argos, Crete,* and *Euboea* dialects.

The Greek letter **PHI** remains unchanged in the above dialects.

The Greek letter **ZETA**(Z) changes to **I** in the above dialects.

## LATIN:

ORIGIN:

Records date back to 1000 B.C.E when migrants from Northern Europe entered the Italian peninsula and settled in a small area called Latium on the banks of the Tiber River. The language derived its name (Latin) from this initial humble settlement of migrants.

It soon spread to the whole of Italy and with the advent of the Roman Empire to the whole of present day East and West Europe.

The Roman Empire ruled the world for about 1500 years At its peak in 117 A.D. exercised control over an area of almost 2 million square miles (approx the size of present day Australia) and almost 90 million people (approx population of present day Germany)!

Classical Latin was the lingua franca of the Roman Empire till 610 A.D following which it was superseded by Greek. In the conquered territories Latin imbibed characteristics of the local inhabitants (Italy, Gaul, and Spain) giving rise to the *Romance* family of languages.

The flowering of the Latin language occurred during the period 80 B.C.E-14 A.D during which time the likes of Cicero, Vergil, Caesar and Ovid flourished.

With the decline of the Roman Empire, the use of Latin as the lingua franca diminished and languages like German, French, and English were increasingly used. Gradually the use of Latin became restricted to the Vatican the seat of the Roman Catholic Church.

## HEBREW:

ORIGIN:

The first evidence of written Hebrew goes back to 3500 B.C.E following the discovery of the *Dead Sea Scrolls* (The Dead Sea is a salt water lake sandwiched between present day Israel and Jordan)

The scrolls themselves, preserved in bottles were found over the period 1946/47-1956 in the Qumran Caves located on the northwestern part of the Dead Sea. The inscriptions were on parchment and papyrus; in some cases on bronze.

The script is in the *Ancient* or *Classical* Hebrew containing verses from the 24 books of the Jewish *Tanach* (Tanach correlates to the 39 books of the Protestant Old Testament and 46 books of the Roman Catholic Old Testament)

Spoken or *Biblical* Hebrew goes back to even earlier times (5-6 millennia) to the Judean *Garden of Eden* and the first humans *Adam & Eve.*

Actually, the use of the word 'Hebrew' to denote a language is of relatively recent origin. Prior to around 1000 B.C.E it was called *the language of Chanaan* and thereafter as *the Jew's language* till around 500 B.C.E.

The first 5 books of the Tanach, Genesis, Exodus, Leviticus, Numbers and Deuteronomy called the *Pentateuch* constitute the *Torah* the Holy Book of the Jews believed to have been written by Moses.

## VOCABULARY & SCRIPT:

Hebrew consists of 22 alphabets; one of these designated by the symbol () has dual sound and hence the number of alphabets in the Hebrew script is stated to be 23. A striking feature of the symbol () is depending on its position in a word the sound generated varies. Some of these resemble vowel like sounds like *w* and *y.*

The English language vowels *a, e, i, o,* and *u* sounds are produced by dots and dashes inserted either above or below consonants.

Hebrew in the Old Testament consist of only 2050 *root words.* Of these only 500 are actually used, restricting the total number of words in the Old Testament to a mere 5000!

This explains inability or difficulty in expressing abstract ideas e.g. In the Old Testament there is no word that means *religion.*

A consequence of this weakness is the use of words in the bible that do not convey the *exact meaning* of what is intended.

Luke 14:26 of the New Testament says:

If any *man* come to me, and hate not his father, and mother, and wife, and children, and brethren, and sisters, yea, and his own life also, he cannot be my disciple (Attributed to Jesus Christ in the KJV of the New Testament)

What was intended could be:

The basic requirement for being my disciple is *single minded dedication and devotion* to my teachings without being distracted by thoughts of the family?

## CONCLUSION:

The objective of this study of languages was to arrive at a conclusion whether it is a God given gift to humans only, excluding all other life forms (Creationist Viewpoint) or is a consequence of Evolution (Evolutionist Viewpoint)?

The only way to answer this question was to go back to the origin of about 6000 languages that are spoken and written in the present day world. This is obviously an impossible task!

Instead we chose 5 of the major languages of the world and looked for clues that *might* lead us to the origin of *all* languages.

This would provide a definitive answer as to the validity of either the Creationist Viewpoint or the alternative Evolutionist Viewpoint.

Since the former goes back to only 6000 years (Adam and Eve as per the Hebraic bible) as against 50,000 years in the latter (Darwinian Neanderthal) we focused on the 5 major world languages mainly because it was more *doable!*

It is obvious that searching for clues from fossil records going back 50,000 years is a near impossible task!

Does it mean that the Evolutionist Viewpoint is *prima facie* invalid in the absence of supporting evidence?

Not necessarily.

We do know that primitive man did acquire a wide spectrum of skills over large swathes of time (thousands of years) to *light a fire, hunt for animal food, farm and harvest food of vegetable origin, and migrate to more habitable areas to escape extreme weather conditions.*

More importantly he was able to *survive and evolve!*

Is not possible therefore that through painstaking trial and error he gradually developed the power of communication through the medium of different languages in different parts of the globe?

In the case of the research into the origin of the 5 chosen languages, going back approx 6000 years (Creationist Viewpoint) we found the period shrouded in antiquity with hardly any evidence that would hold up to modern scientific scrutiny.

Like many other observed present day phenomena, there is no definitive answer as to how exactly humans acquired the skill to communicate with each other through the medium of language!

**Illustration 12:** The Tower of Babel- Shutterstock _1817928545

# Seventeen

# Does God Exist?

Throughout history going back more than 5 millennia, humankind has wrestled with the concept of GOD.

Even today, we do not have an answer to the eternal question: DOES GOD EXIST?

The answer is illusory and linked to the belief systems of the world's major religions.

There is also the 'Unaffiliated Group' a motley collection of people who are unwilling to be associated with any world religion. Atheists, agnostics, and scientists with an alternative well reasoned version of God i.e. the maker of immutable laws governing the working of the universe.

As the third or fourth largest group, the 'Unaffiliated Group' trails Christianity, Islam and Hinduism. its significance in defining the world religious cannot be downplayed.

Pew Research Centre, Washington, USA estimates a 15% share of world population adhering to the Unaffiliated Group at present and future projections till 2050!

Most of us tacitly believe in the God/Religion of our birth with the accompanying traditions and practices.

Notwithstanding this, there are considerable regional variations in the intensity of belief and observance of religious practices and rituals particularly amongst Hindus.

Uniformity and a high degree of compliance is achieved in Islam, where majority of Muslims congregate on Fridays (sacred day) for prayer.

On this day mosques throughout the Muslim world publicly exhort believers five times ('azan' ) to participate in the prayer sessions.

Likewise, most Christians attend the Sunday prayer session in the community church they are traditionally attached to.

Throughout recorded history there are no instances of God and miracles holding up to verifiable scientific scrutiny. Despite this, more than 80% of the world population believe in God either as a concept or in some concrete shape and form!

Culturally, irrespective of the religion one follows, it is the standard norm for humans to have faith in their chosen God and Religion and observe its practices.

Statements like:

*I DO NOT BELIEVE IN GOD* in public would be reacted to with anger, consternation and even pity for the speaker by the majority of people gathered around.

Considering all religions that exist today, the different dogmas, shades of conviction and belief in the God concept, religious festivals, religious practices and rituals is indeed mind boggling!

Scientists, philosophers, historians, and religious scholars have struggled for centuries to find words that can convey subtle variations in meaning relating to the God concept. Some examples:

## IMMANENCE:

Something intrinsic, innate and omnipresent e.g. 'God is immanent in the world'

## ONTOLOGY:

Everything we see in the world, both animate and inanimate.

## EPISTEMOLOGY:

Study of our knowledge of nature relating in particular to its limitations and validity.

## MONISM:

Theory that all phenomena in the world is caused by a 'single' principle that can be called 'God'

## DUALISM:

Humans consist of two basic elements- matter and spirit that can be linked to the God concept.

We have tried to explain these words in the context of the God concept.

Neither the list nor the meanings of words that have been used to define God are unambiguous or comprehensive.

This is not surprising since we are dealing with a topic that has had an air of mystery surrounding it from the earliest of times.

Such thought processes are deeply interlinked with historical events like the emergence of 'Gods' 'godmen' and 'godwomen', miracles and events recorded in religious texts that border on the 'supernatural'.

Before elaborating further, it would be useful to define keywords connected with God and religion for a better perspective:

## I THEISM

Theism is a convenient starting point in explaining a few related God/Religion terms that will throw light on the manner in which this concept evolved in the long history of humankind.

We rewind our clocks to the Old Testament of the Hebrew biblie the starting point of the monotheistic religions-Judaism, Christianity, and Islam.

# II 'PERSONAL GOD' - THE CREATIONIST

I am outlining the narrative from the point of view of Judaism and Christianity that are the earliest religions.

Christianity has the largest number of adherents today (32% of world population).

Judaism believed in Yahweh and Christianity in the Father in Heaven and His Son Jesus Christ.

These constitute the earliest *Personal God* adherents.

Six centuries after the commencement of the Christian Era, the world witnessed the entry of Islam with the 'Personal God' Allah- and the Holy Word of the Koran revealed to the last of the prophets Muhammad

This completes the trinity of 'Personal Gods'

We shall probe further into the Personal God of Christianity the most widely practised religion in the world today.

We wind our clocks back to the hoary days of Adam and Eve (Old Testament of the Hebrew bible) who lived an idyllic life in the Garden of Eden. They were made in 'God's own image' and enjoyed a status superior to all other living beings on land, sea, and the air.

There was a restriction placed on them i.e. they should not eat the fruit of the 'Tree of Knowledge of Good and Evil'

We do not know why a benevolent God imposed such a restriction.

The name of the forbidden tree i.e. Tree of Knowledge of 'Good' and 'Evil' gives a clue. Everything was good till the fruit was eaten since the tree besides being 'good' also had an 'evil' element which God wished to protect humankind from?

The devil was determined to break the sacred bond between God and Man and successfully tempted Adam and Eve to violate God's edict.

This is the 'Taint of Original Sin' which affected Adam and Eve and all their descendants thereafter.

How can Man overcome this impediment and become acceptable to God the Father?

The solution came millennia later through the birth of Jesus Christ, the Son of God. His sacrifice (death by crucifixion) on the Cross redeemed believers of their 'Original Sin'.

On The Day of Judgment (Second coming of Jesus Christ) they would be saved and join the Father in Heaven for eternity.

Implicit in the 'Personal God' belief system is the existence of a 'Creator' who brought our Universe and everything in it into existence-the galaxies, stars, the solar system, humans, plants, animals,avian, and marine life.

A related belief is the 'Anthropomorphic Concept' that 'God made Man in his own Image' and granted him dominance over all living creatures on land, in the air, and in the sea!

These are articles of faith in Judaism and Christianity, inextricably linked to each other through prophecies in the Old Testament being fulfilled in the New Testament.

## THE LOGIC OF BELIEF IN GOD/RELIGION

At the commencement of the history of Judaism initially, and Christianity later, the motivation for acceptance of the God concept (Personal God) arose mainly out of fear and an innate desire of humans of that era for a 'Big Brother'.

This fulfilled their need for protection from natural calamities, ailments, enemies and forces of nature like lightning, thunder, weather changes,the sun, moon, stars, and eclipses that they could neither understand nor control.

This allegiance came at a cost- there had to be absolute faith in Yahweh (Judaism) and the Father (Christianity). In the latter case the faith had to extend to the Son (Jesus Christ) and inter alia to the belief that the death of Jesus by crucifixion would redeem believers from the taint of 'Original Sin'

Belief further extends to the Day of Judgement when humankind would be 'resurrected', believers rewarded and non believers punished for eternity!

We shall now comment briefly on Islam the second largest practised religion in the world:

Six to seven centuries into the Christian Era, Islam made its entry through the Word of God being revealed by Archangel Gabriel to Muhammad. This happened over a period of thirty years initially in Mecca and later in Medina both in present day Saudi Arabia.

Islam is the third and last monotheistic religion with unique teachings contained in the Koran believed by Muslims the world over.

Like Christianity Islam believed in the Day of Judgement and the resurrection of humankind for facing the trial involving the observance during the individual's lifetime of the Islamic Sharia law.

Also the faith that Allah is the only God whose true teaching (Koran) was revealed to the world through the final prophet Muhammad!

Asian religions like Hinduism, Buddhism, Sikhism, Jainism follow a different track. There is no adherence to a Personal God, Original Sin, Day of Judgement, Heaven and Hell.

Instead an individual's life is determined by his/her 'samskara', 'karma' (Hinduism) 'nirvana' (Buddhism) and similar 'esoteric' concepts in major religions/philosophies like Sikhism, Jainism, Taoism, Confucianism!

## DEISM AND THE GOD OF SPINOZA

These two depictions of God are similar. God is considered as a 'clockwinder' of the universe and 'absent' before, during, and after the lifespan of humans and other lifeforms on earth.

Events that unfold including events in an individual's life are the result of the inexorable 'unwinding' of the universal clock.

Events are 'fixed' i.e. there is no 'free will' for either humans or anybody else!

The 'Anthropomorphic Concept' of God making 'man in his own image' is rejected.

Deism and the God of Spinoza and variations of it have many adherents in the scientific community like Einstein, Weinberg, and Hawking.

## PALEONTOLOGY- THE EVOLUTIONIST

Much later, paleontologists based on a Darwinian study of fossil remains reconstructed history upto our earliest ancestor the apelike Neanderthal Man estimated to have lived on earth 50,000 years ago.

If you take the biblical Adam as the first man (going back 6000 years), there is obviously a large swathe of time separating Adam from the Neanderthal Man.

This difference has not been satisfactorily reconciled to this day.

The schism between the Creationists (Believers in a Personal God) and Evolutionists (Our Universe exists because of a 'lucky fluke') at the 'beginning' is still very much in vogue today.

An example of the lucky fluke from our day to day experience is the dealing of a Royal flush (AKQJ10 of the same suit) to a poker player! Such an event is 'highly unlikely' but NOT 'impossible'

Eminent scientists and philosophers are listed in both camps!

In the Creationist Camp we have famous physicists like Isaac Newton, Rene Descartes Stephen Barr, and surprisingly Charles Darwin the Founder of the theory of Evolution!

The Evolutionist Camp boasts of members like Albert Einstein, Stephen Hawking. The group rejected the 'Personal God' concept but accepted the Spinoza version of God i.e. a God who laid down immutable laws governing nature's working.

## THE UNAFILLIATED: ATHEISTS AND AGNOSTICS:

Religions offer solutions based on faith, but these are not acceptable to the science purists many of whom are either Atheists or Agnostics.

Instances of scientists in the Atheist group include Steven Weinberg (arguably the world's leading particle physicist) and Richard Dawkins (famous biologist)

Steven Weinberg says:

'With or without religion there will be good people doing good things, and evil people doing evil things, but it takes religion to make good people do evil things' and in another context:

'I dislike religion intensely'

American astro physicist Neil degrasse Tyson describes himself as an agnostic.

## THE PHYSICS VIEWPOINT:

Theoretical physicists established a timeline for Our Universe (Big Bang-14 billion years ago) followed by birth of our solar system (4.5 billion years ago

How does modern science view the God concept?

In the 20th and 21st century, particle physicists coined the word 'God Particle' a subatomic particle existence of which was necessary to answer questions in Quantum Physics. The particle was predicted by British theoretical physicist Peter Higgs (1929- ) in 1960.

The particle itself was discovered 52 years later in 2012 by experiments conducted in the Large Hadron Collider in Europe.

In recognition, Higgs was awarded the Nobel Prize for physics along with French physicist Francois Eglert in 2013.

The God Particle was renamed as the 'Higgs boson', a subatomic particle believed to be the 'ultimate ingredient' providing 'mass' to all matter.

The coexistence of Religion and Science in respect of the God concept is essential for the peace and happiness of humankind.

The fundamental problem is based on the scientist's inability to accept the notion of a 'Personal God' as understood and accepted without question in the three monotheistic religions, Judaism, Christianity and Islam.

Scientists do accept the 'God of Spinoza' the Dutch philosopher. As already stated this belief system is based on immutable laws that govern every aspect of the working of our universe including the evolution of humans and the millions of species of animal, avian, marine, and plant life.

All scientists do not fall into this group of 'evolutionists'. There are scientists who are staunch followers of religion, the 'creationists'.

As a matter of fact the creationists outnumber the evolutionists!

And to complete the tally, there are followers of the 'esoteric religions' like Hinduism, Buddhism, and Confucianism that have a philosophical outlook on life.

Nowhere is this schism better depicted than in the case of Steven Weinberg (arguably the leading particle physicist in the world today) an atheist and Abdus Salam a staunch follower of Islam.

Weinberg and Salam were jointly awarded the Nobel Prize for physics in 1979 for their theory unifying the weak nuclear and electromagnetic fundamental forces.

The third recipient who shared the prize was Sheldon Lee Glashow.

Sadly, the resolution of the religion science dichotomy is still an ongoing quest; the desired target of total convergence of the two is still a long way off.

# PART II

## MATH CONUNDRUMS

# Eighteen

# Mathematics ~ A Unique Science

*Mathematics may not teach us to add love or minus hate. But it gives us every reason to hope that every problem has a solution.*

**Anonymous**

Mathematics is delightful to those who know how to unveil its beauty. It enables us to unravel the mysteries of the universe by partnering with its senior cousin *physics*. Aside from this role, it serves as a building block for software, architecture, engineering, art, finance, weather predictions, cryptography… the list is endless.

There are many misconceptions about mathematics. A few are listed below:

- *Provides limited job opportunities.*

- *Mathematicians are socially unacceptable.*

- *Mathematicians are mere statisticians*

- *Math students are prone to commit suicide.*

A little thought shows that the above commonplace sentiments in the public mind are baseless and not even worthy of being refuted. There are more meaningful thoughts on mathematics that are discussed below:

- At job interviews a question often thrown at a candidate with a math background is

*Can you perform calculations in your mind faster than others?*

The query arises out of the erroneous belief that math geniuses are human calculators. Most are *not*.

- They are adept in probing *mathematical structures,* formulating *theorems* and resolving *riders* based on an understanding of the theorem.

*What are Mathematical Structures?*

These could be a *set endowed with special features* like an arithmetic sequence,

1, 4, 7, 10, 13, 16, 19, 22, 25,…..

The difference between one term and the next is constant - in the above instance it is 3. In general we could express the sum total of the set symbolically as

a, a+d, a+2d, a+3d, …..or

$x_n$ = a+ d(n -1) which when applied to the second term in the set

= 1 + 3(2 -1)

= 4

as per the equation.

Suppose you wish to derive the 9th term using the above equation:

$X_9$ = 1+3(9 - 1)

= 1+3(8)

= 25 which is correct.

We can derive the equation that gives us the sum(S) of the terms in an arithmetic sequence as follows:

S = a + (a+d)+……….+(a+(n-2)d) +(a+(n-1)d

We next rewrite S in the reverse order:

S = (a+(n-1)d) + (a+(n-2)d) +…..+(a+d)+a

If we now add the two, *term by term,* and divide the end result by 2 the result would be:

$S_n$ = (n/2) x (2a+(n-1)d)

which we call an *arithmetic series or arithmetic progression.*

Let us check out the equation for finding out the sum of the first 9 terms of our earlier sequence 1,4,7,10,13,16,19,22, 25:

$S_9$ = (9/2) x (2+(9-1)3)

   = (9/2) x 26

   = 234/2

   = 117 which is correct.

The reason for the success of this formula is:

1+25 =26 first plus last

4+22 =26 second plus second last

7+19 =26 third plus third last

.................

.................

.............. till $n$ rows are completed.

The mathematician who gifted this finding to the world was Carl Friedrich Gauss, German mathematician of the 19th century.

There are other sequence types like *geometric, quadratic, harmonic, special* and *Fibonacci numbers.* Fibonacci numbers come up for special attention later in our book. We are not explaining the others as it would be an unnecessary digression from the main theme.

- *Theorems and Riders:*

The starting point is an *axiom*(rule assumed to be true) from which we build *theorems* that have to be proved rigorously. The truth in an axiom is *self evident* and does not require proof e.g. the axiom,

*If equals be added to equals, the wholes are equal.*

Or another well known axiom,

*The whole is greater than the part.*

The best known example of a theorem is the Pythagoras Theorem stating that the square on the hypotenuse of a right angled triangle is equal to the sum of the squares on the other two sides.

A good example of a random *rider* (in this case not based on the Pythagoras Theorem) relates to -

a point within an *equilateral triangle* ABC from which perpendiculars $p_1 p_2$ and $p_3$ are dropped to the three sides AB, BC, and CA. Next perpendicular $p$ is dropped from any of the vertices, A, B, or C to the opposite side. The rider requires you to prove that

$$p = p_1 + p_2 + p_3$$

(The solution is easy. I can assist the reader if needed)

**Mathematics as a language:*

Italian astronomer *Galileo Galilei* is reported to have said - *Mathematics is the language in which God has written the universe.*

Galileo's reported statement is debatable; however when two mathematicians converse with each other it is difficult for a *historian* to participate meaningfully. Mathematics is in this sense a different language like Cantonese or Swahili. In normal parlance when we speak of a language we mean a vehicle that enables us to describe in words and/or sentences an object, an emotion or feeling that we wish to convey to others. For example to describe the colours of a rainbow we would say violet, indigo, blue, green, orange and red. If we try to translate the concept to the *language of mathematics* we would first need to define violet, indigo…and create symbols for them for future use when they *interact* with one another.

Another difference is that mathematics is *exact* whereas a conventional language like Tamil or Mandarin has words that can convey different *shades of meaning.*

Relationships in math language are *self consistent.* Conventional language statements are *arbitrary* and not required to be *self consistent.*

Mathematics as normally used is *not* a language. It can be quite *unfriendly and lifeless.* The Pythagoras Theorem we spoke of earlier would be expressed as

$c^2 = a^2 + b^2$ where *a, b,* and *c* are the two sides and hypotenuse respectively of the right angled triangle ABC.

We conclude this introduction to the fascinating subject of mathematics with an appeal to readers *not* to get intimidated by Greek symbols like $\Sigma$ - *Sigma* that indicates *the sum of e.g.* $\Sigma X_n$ *is an abbreviation for* $X_1 + X_2 + X_3 \ldots$

$\int$ *is the integral sign used in integral calculus,* $\infty$ *indicates infinity, and* $\partial$ *(delta) is used in differential calculus* to name a few. Rather regard mathematics as a friend, *an enabler* and *predictor* to be used to develop new avenues of knowledge yet to be discovered.

I have handpicked a few conundrums that are special. There are thousands of similar jaw dropping problems for the math enthusiast in the public domain that can be explored.

# Nineteen

# 666 ~ The Number Of The Beast

*Here is wisdom. Let him that hath understanding count the number of the beast: for it is the number of a man; and his number is six hundred and sixty six*

**New Testament Hebraic Bible Revelation 13:18**

1. *Angel Numbers*

The belief in some circles that the number 666 is unlucky and associated with the devil is not justified. This misconception had arisen 2 millennia back covering Judaism and a period in Christianity that followed the death of Jesus Christ. It got reinforced in recent times through movies like *Omen I* where the child *Damien* possessed by the devil is seen to have the scar of 666 on his skull. On the contrary, numbers with repeating digits like *11: 11* (eleven minutes past eleven) or car number plates ending *2222 / 3333* are considered to be *significant spiritual messages* and called *Angel Numbers*. If anybody is continuously noticing such numbers it cannot be dismissed as a coincidence. 666 in particular is a call from above to a person to pay closer attention to any fixation he or she may have on an earthly problem. It can be considered as a *divine nudge* to redirect energy. A clue on how to proceed is provided by the *location* and *timing* when the angel number is perceived. If it is frequently happening in the context of an individual it could indicate that you are overlooking the value of the relationship over a relatively trivial development. Striving for *perfection* is futile. Have faith in the universe to see you through difficult situations.

- *Magic Square*

666 can form a 6 X 6 magic square with extraordinary properties as displayed below:

*6 32 3 34 35 1* 6+32+3+34+35+1=**111**

*7 11 27 28 8 8 30* 7+11+27+28+8+30=**111**

*19 14 16 15 23 24*

19+14+16+15+23+24=**111**

*18 20 22 21 17 13*

18+20+22+21+17+13=**111**

*25 29 10 9 26 12*

25+29+10+9+26+12=**111**

*36 5 33 4 2 31*

36+5+33+4+2+31=**111**

TOTAL **666**

Sum total of arithmetic series:

1 + 2 + 3 +……... 36=**666**

as per formula: $36(36+1)/2 = 36^2+36/2 = $**666**

*Other noteworthy features:*

# The numbers 1,3,6,10,15,21,28,36, form a *Triangular Number Sequence*. They represent the number of dots in an imaginary equilateral triangle. For example the triangle with 6 dots is sketched below:

```
            *

        *       *

    *       *       *
```

The first triangle has only one dot.

The second triangle has another row with 2 extra dots, making 1 + 2 = 3

………….

…………..

………….

The *nth* triangle would have $x_n = n(n+1)/2$.

Let us check it out for the 4th triangle we had drawn a sketch of:

$X_4$ = 4(4+1) / 2 = 10 which is correct. Suppose you wish to calculate the number of dots in the 50th triangle:

$X_{50}$ = 50(50+1) / 2 = 1275 dots

# The numbers 4, 22, 27 in our magic square and 666 are *Smith numbers.*

A Smith number is a composite number which in the normal decimal base 10(or any other well defined base), the sum of its digits is equal to the prime numbers that comprise it. For example take the number 27 that is equal to 3 x3 x 3. It will be observed that 3 + 3 + 3 = 2 + 7 = 9

The Angel Number 666 is *also* a Smith number because:

666 = 2 x 3 x 3 x 37 and

6 + 6 + 6 = 18 = 2 + 3 + 3 + 3 + 7 = 18.

# Another interesting feature of 666 is that it is the total of the first 144 digits that follow the decimal point in $\pi$(Pi) i.e.

3.141592653358979932384……..till 144 digits.

| 6 | 32 | 3 | 34 | 35 | 1 |
|---|----|---|----|----|---|
| 7 | 11 | 27 | 28 | 8 | 30 |
| 19 | 14 | 16 | 15 | 23 | 24 |
| 18 | 20 | 22 | 21 | 17 | 13 |
| 25 | 29 | 10 | 9 | 26 | 12 |
| 36 | 5 | 33 | 4 | 2 | 31 |

6 +32+ 3 +34+35+ 1 = **111**

7 +11+27+28+ 8 +30= **111**

19+14+16+15+23+24= **111**

18+20+22+21+17+13= **111**

25+29+10+ 9 +26+12= **111**

36+ 5 +33+ 4 + 2 +31= **111**

**666**

1 + 2 + 3 + ... + 36 = **666**     $\dfrac{36(36+1)}{2} = \dfrac{36^2 + 36}{2} = 666$

**Illustration 13:** The Number of the Beast - 666- Shutterstock_2120756993

# Twenty

# Perennial Mystery With Numbers

*Numbers do not feel. Do not bleed or weep or hope. They do not know bravery, sacrifice, love and allegiance. At the very apex of callousness, you will find only ones and zeros.*

**Arnie Kaufman (Australian author of science fiction.
Presently resides in Melbourne)**

Figure 1

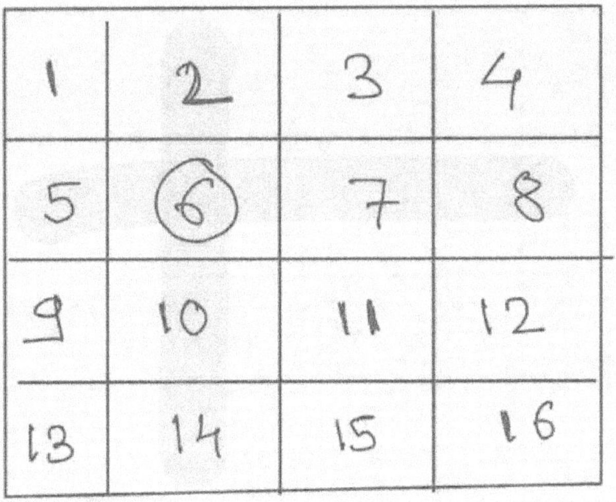

| 1 | 2 | 3 | 4 |
|---|---|---|---|
| 5 | (6) | 7 | 8 |
| 9 | 10 | 11 | 12 |
| 13 | 14 | 15 | 16 |

Consider the innocuous looking number square in Figure 1

A casual glance reveals nothing more than a 4x4 square with numbers 1-16 in 4 rows arranged in ascending order.

Now attempt this mental exercise.

Choose any one number in the box and commit it to your memory. Pick a second number *from a different row and column* of the first number. Assume you had picked 6, then the second number should *not* be from the row and column highlighted in Figure 2 below.

You need to commit numbers 1 and 2 to memory or write them down before proceeding further.

Next, pick a third and fourth number with the same condition i.e. they should not fall in the rows and columns of the first two numbers in Figure 2.

These actions are graphically displayed in Figures 2 and 3 given below

Figure 2

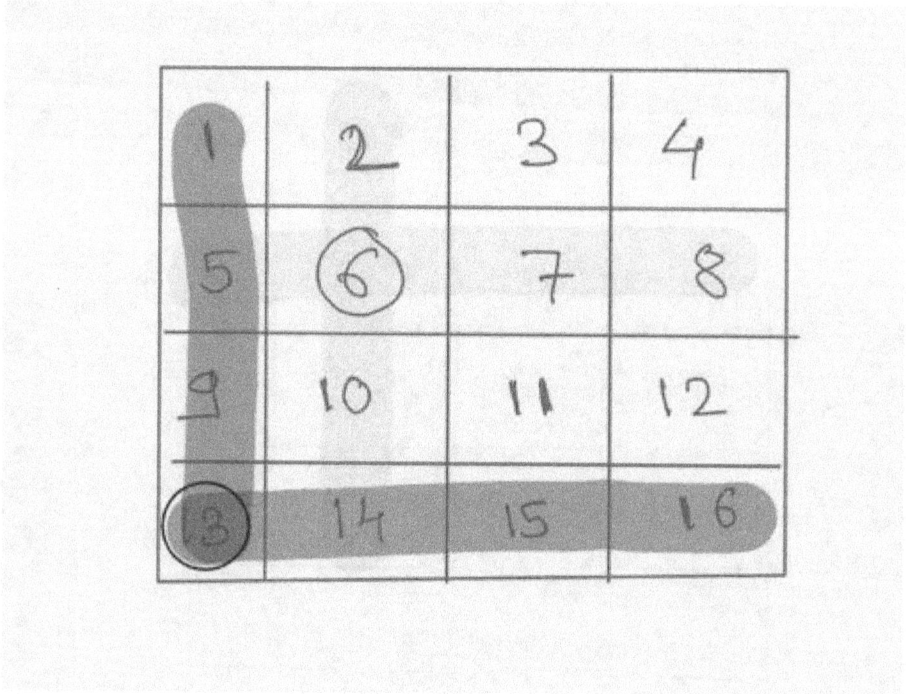

## Figure 3

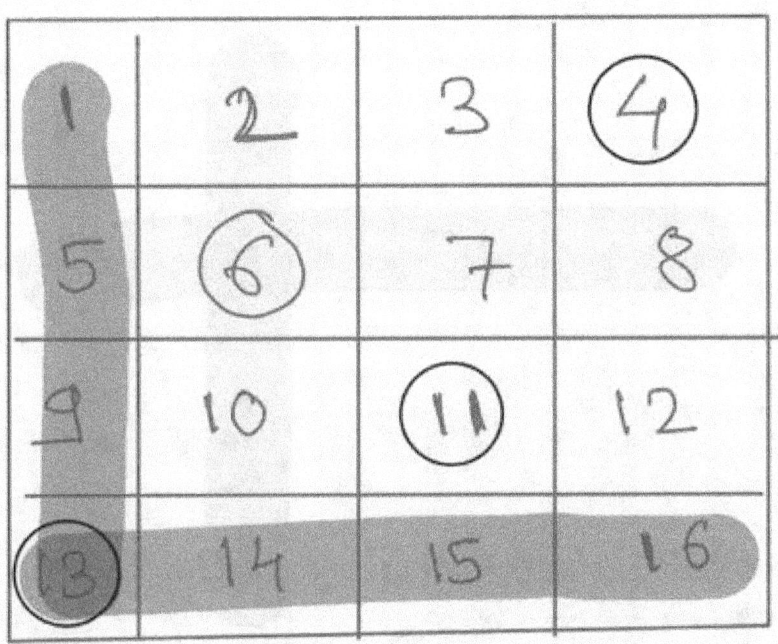



The 4 chosen numbers are nominated as A, B, C and D.

Next calculate the total:

A + B + C + D

The end result (Figure 4) is always the same irrespective of the chosen numbers provided the rules have been followed.

Figure 4

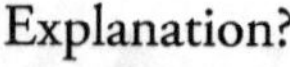

Explanation?

The 4x4 box is called a *matrix*.

The sum of the numbers in the 2 diagonals is the *trace* of the matrix.

If you pick any 4 numbers in the box provided that there is no *overlapping of rows and columns* in the 4 chosen numbers, the trace will always be 34.

Similarly for a 2x2 matrix the trace is 5,

3x3 matrix 15, 5x5 matrix 65.....

*Can you figure out the trace for a 6x6 matrix?*

# Twenty One

# The Power Of The Exponent

*The only way to learn mathematics is to do mathematics.*

**Paul Richard Halmos (1916 – 2006)**
**Hungarian born American Mathematician.**

There is an interesting question that exemplifies the above concept.

Which is greater $99^{100}$ or $100^{99}$

We reformat the question to read as,

$99^{(1+99)}$ or $(1+99)^{99}$

To generalize, we state,

$(1+n)^n$ to be compared to $n^{(1+n)}$ and work out the result for natural numbers n = 1,2,3….99 to observe a possible trend as n increases and extrapolate the end result.

For example, for n =1, we get $(1+1)^1$ equal to $2^1 = 2$ as compared to $1^{(1+1)}$ or $1^2 = 1$.

n=2, we get $(1+2)^2 = 3^2 = 9$, and $2^{(1+2)} = 2^3 = 8$.

n=3, we get $(1+3)^3 = 4^3 = 64$, and $3^{(1+3)} = 3^4 = 81$.

n=4, we get $(1+4)^4 = 5^4 = 625$, and $4^{(1+4)} = 4^5 = 1024$.

n=5, we get $(1+5)^5 = 6^5 = 7776$, and $5^{(1+5)} = 5^6 = 15625$.

The trend is clear. For n>2, $(1+n)^n$ is less than $n^{(1+n)}$ and thereafter the differences increase rapidly.

We therefore conclude that $99^{100}$ is greater than $100^{99}$.

It demonstrates the awesome power of the exponent!

# Twenty Two

# The Origin Of Zero

*Mathematics is a more powerful instrument of knowledge than any other that has been bequeathed to us by human agency.*

**Rene Descartes (1596 – 1650)**

Rene Descartes was a French philosopher, mathematician, and scientist who linked algebra and geometry to create analytic geometry.

We often overlook the pivotal role played by the humble *zero* in the evolution of mathematics. This ancient development is forgotten in the euphoria of modern technology.

The zero is the basis of the *binary code* that saturates the world of calculations.

Recently published findings by the University of Oxford pushed back the first recorded use of zero to a300 C.E. Indian manuscript. This finding is based on the carbon dating of an ancient mathematical treatise the *Bakshila Manuscript*.

Some researchers consider this apocryphal and nominate a more realistic date for the discovery of zero to the era of the formidable Indian mathematician foursome of Brahmgupta, Mahavira, Bhaskara I and Bhaskara II about 3 centuries later i.e. 600 - 700 C.E.

In short, experts agree that zero is an Indian invention but the date of commencement of its usage differs in the narratives followed in the ancient cultures of *Babylon, Greece and Rome*.

Zero makes shadowy appearances in historical narratives only to vanish each time into temporary oblivion.

It is almost as if the mathematicians of yore were searching for signals that would throw more light on the evolution of zero and yet did *not* recognize them when they actually appeared!

Be that as it may, from a global perspective, the concept of zero meaning *nothing* did exist elsewhere before any written evidence emerged from the Indian subcontinent.

In mathematics, the distinction between zero as a *place holder* and *numeral* is significant e.g. in the number 9807 the zero is used so that the positions of 9 and 8 are correctly identified. Clearly 987 is quite different from 9807, zero playing the important role of a placeholder.

The second use is as a numeral meaning *nothing or no value.*

Neither of the above uses has an easily described history. As is normally the case with inventions and discoveries, it is not as if someone invented these ideas together or separately on a particular date and everyone started using them ever since.

Today everything relating to zero seems obvious and self explanatory, yet the *Babylonians* were happy using a number system without zero for over 1000 years. (Babylon is a name that figures prominently when probing the history of numbers. This ancient town is located along the Euphrates River in present day Iraq, 50 miles south of Baghdad. It was destroyed in 539 BCE by the Persian Emperor Cyrus the Great. The biblical *Tower of Babel* was located in Babylon but no longer exists)

The history of number evolution is likewise prominent in ancient *Greece.* Eminent Greek mathematician Euclid 300 BCE wrote a book on number theory called *Elements* but surprisingly, it is based *not* on natural numbers but on geometry.

Amongst other early contributors to the number theory we could include *the Almagest* written by *Claudius Ptolemy* 100 CE the noted Roman mathematician and astronomer.

The Indian contribution which came several centuries later needs to be understood in the context of the history of mathematics pertaining to ancient times across the globe.

Its significance should neither be exaggerated nor underplayed.

The first record of the Indian use of zero agreed by all is 876 CE based on discovery by archaeologists of a stone inscription in Gwalior, 400 kms south of Delhi in which Indian mathematicians like Brahmgupta are believed to have played a role.

Indian mathematicians of that era struggled with the concepts of addition, subtraction, multiplication and division. Apart from the different nuances of zero they had a partial understanding of the consequences of dividing by zero, i.e. peeking into *infinity.*

Strangely, over the centuries that followed right up to the present day problems posed by zero have not come to an end. For example, modern computers that are subject to constant innovation run on a binary - zeros and ones.

The human race's discovery of zero was a *total game changer.. equivalent to learning a new language* says Andreas Nieder, Cognitive Scientist at the University of Tubingen, Germany.

Other life forms like monkeys and bees have brains that equate zero with no value. Humans are the sole exception who seek to use it as a tool.

We regard the absence of something as *a thing in itself.* Even in the darkest emptiness of space there is always *something, a true zero.*

The brilliant American mathematician of Hungarian origin, John Von Neumann described an empty box as a box with the *empty set.* If you place another empty box inside the first, a third inside the second ....we are building a *set* and simultaneously teaching ourselves how to count one, two, three, four .......

*It is the nothing that is.*

Zero stands as the far horizon beckoning us the way *horizons do in paintings. It unifies the entire picture. If you look at zero you see nothing. But if you look through it you see the world. It is the horizon.*

More significantly, without zero there would be no negative or imaginary numbers, and above all the weird concept of *infinity* would cease to exist and with it *calculus* and its manifold applications.

The concept of zero is *not* hard wired into the human life form like a language. We have to learn it, and it takes time.

A child younger than six has difficulty in understanding which number is smaller one or zero. Most children under six if questioned would opt for one!

The subject seems deceptively simple. In reality it is not so. It forms an integral part of of a whole field of study termed *Cognitive Sciences.*

In view of the relevance of this topic to our present essay we are providing a summary of the points made and seek the reader's patience to bear with occasional repetition of certain concepts.

We start with the distinction between zero as a placeholder and a numeral:

Placeholder zeros, according to Harvard math researcher Robert Kaplan, were first documented 5000 years ago in Mesopotamia by the Sumerians.

Similarly, the Chinese who used sticks for counting and the Babylonians were aware of zero as a placeholder.

India was the first location where it's significance as a numeral was understood.

It was used here in equations by *Vedic mathematicians* according to Peter Gobets a leading member of *Project Zero* a collection of academics who are trying to determine the origin of zero.

*Indian mathematicians were able to conceive and fully invest in it because of their philosophical understanding of* **shunya** *the nothingness that is the natural counterpart to something.*

It is difficult to overstate the importance of this. Mathematics of those times was more an expression of philosophical ideas and reasoning, directed towards abstract studies like astronomy and *not* commerce.

"Today we take it for granted that the concept of zero is used across the globe and is a key building block of the digital world. But the creation of zero as a number in its own right, which evolved from the placeholder dot symbol found in the Bakhshali manuscript, was one of the greatest breakthroughs in the history of mathematics" says *Marcus du Sautoy,* professor of mathematics at the University of Oxford.

We now believe that mathematicians in India planted the seed of the idea that zero is a numeral that would later become so fundamental to the modern world.

This is also the explanation why zero as a numeral could *never* have been conceptualized by the foremost thinkers in the West e.g.the Greeks who abhorred the idea of nothingness.

Pre Socratic schools of thought held the view that

*"Nothing cannot be something"*

This stand was confirmed by Aristotle a couple of centuries later.

In contrast, contemporary India proved a fertile ground for the evolution of zero from a placeholder to a numeral. The concept of of *shunya* is embraced completely, and is explored as being complementary to the concept of fullness. It is not just math, or astronomy that it influences, but due to Vedic influence extends to language, dance and music. Indians actually utilized zero in calculations thereby elevating it from placeholder to numeral.

The Bakhshila manuscript was unearthed in *Mardan* in present day Pakistan. Indologist *AFR Hoernle* took it to the *Bodleian* Library in Oxford.

Japanese scholar *Dr.Hayashi Takao* placed its age between $8^{th}$ and $12^{th}$ centuries. Oxford's research based carbon dating favours an earlier date making it the *first recorded use of zero by the civilization that made it a numeral.* Researcher Gobets regrets the paucity of evidence in conducting investigations. Most of the pertinent documents were scripted on perishable birch bark stored without long term preservation in mind. The precise date when zero acquired number status is still contested but the fact that it spread from Sanskrit treatises in astronomy to the rest of the world is generally accepted. The long checkered history of the humble zero over the past 2 millennia is exciting, fascinating and barely credible.

# Twenty Three

# Euler's Identity ~ The Most Beautiful Equation

*Mathematics is not about numbers, equations, computations, or algorithms. It is about **understanding**.*

**William Paul Thurston (1946 – 2012)**
**American mathematician specializing in topology.**

The Swiss mathematician Leonhard Euler (1707 – 1783) is acclaimed as the *king of mathematics*. He founded graph theory and topology besides making path breaking discoveries in analytic geometry, trigonometry, calculus and number theory. He was totally blind for the last 12 years of his life. His memory was so remarkable that almost half his total works were produced during this period despite the handicap. He is considered by many to be the *most prolific writer of mathematics of all time*.

The most beautiful equation, called Euler's Identity is:

$$e^{i\pi} + 1 = 0$$

In the mathematics world it is just one step behind Einstein's celebrated equation on Special Relativity,

$$E = mc^2$$

For the mathematical purist it is difficult to separate these two equations from the angle of their significance in the world of science.

Be that as it may, the Euler's Identity in math, also referred to as *Euler's Formula* has been compared to a Shakespearean sonnet. American theoretical physicist Richard Feynman called it a *jewel* and *the most remarkable formula in mathematics.*

What is special in the Euler Identity?

*It comprises the five most important mathematical constants:*

- The number 0

- The number 1

- The number $\pi$ an irrational number

- The number $e$ the base of natural logarithms, also an irrational number.

- The number $i$ or the negative square root of 1, the fundamental imaginary number.

The creator of the identity, Leonhard Euler with an unparalleled output in mathematical discoveries most of which emerged in the last two decades of his life when he was totally blind.

Euler's Identity stems from interactions of *complex numbers,* comprising a real part and and an imaginary part

e.g. $5 + 4i$ used extensively in electrical engineering, quantum mechanics and many other subjects like computer graphics, robotics and flight dynamics.

Computation with complex numbers causes the phenomena of *rotation* and *dilation* that are complicated topics, details of which are beyond the scope of this essay.

Euler's importance arises because it establishes the fundamental relationship between *trigonometric* and *exponential* functions. From a geometry viewpoint it is a way of bridging two representations of the *same unit complex number in the complex plane.*

It shows a most profound connection between fundamental numbers in mathematics - 0, 1, $e$, $i$ and $\pi$.

In addition it proves that $\pi$ is *transcendental* and as a consequence it is impossible to *square a circle.*

Much of the above is beyond the scope of a reader without a college level knowledge of mathematics.

Despite his profound capability in mathematics, Euler was a staunch believer in God and wrote on *the nature of God and the human spirit!*

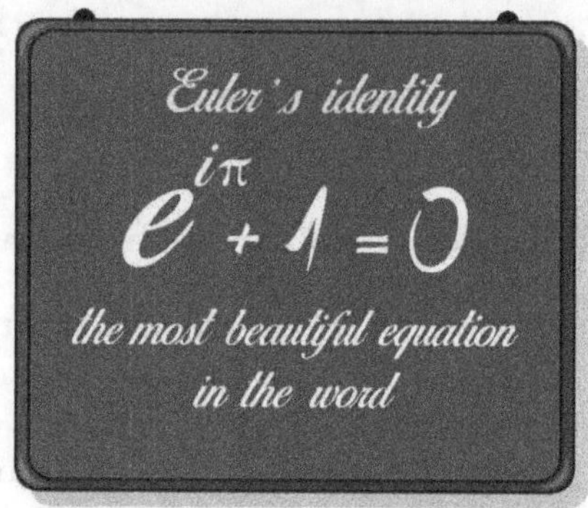


**Illustration 14:** The Most Beautiful Equation - Euler's Identity
- Shutterstock _ 1672085371

# Twenty Four

# Chaos Theory

*Do not worry about your problems in mathematics. I assure you, my problems with mathematics are much greater than yours. God does not care about our mathematical difficulties.*

**Albert Einstein (1879 -1955)**

*The coexistence of chaos and orderliness*

The subject deals with complex workings of diverse natural systems as the beating of the human heart and the trajectory of asteroids. At the centre of *Chaos Theory* is the fascinating idea that order and chaos are *not* always diametrically opposed. Chaotic systems are an intimate mix of the two - from the outside they display unpredictable and chaotic behaviour, but expose the inner workings and you discover a perfectly deterministic set of equations *ticking like a clock.*

Some systems flip this premise around, with orderly effects emerging out of turbulent and chaotic causes.

How can order on a small scale produce chaos on a larger scale?

And how can we tell the difference between pure randomness and orderly patterns that are cloaked in chaos?

The answers can be found in three common features shared by most chaotic systems. Let us look at some examples in our day to day lives:

Jeff Goldblum of *Jurassic Park* fame makes this profound statement:

*Tiny variations…never repeat, and vastly affect the outcome.*

Take the case of the butterfly. A small innocuous creature that flips around flowers creating happiness in those who are lucky to watch it. It is an example of *fragile delicate beauty* not of strength and power. Yet in weather forecasting meteorologists have discovered that the *flapping of a butterfly's wings could cause a hurricane half a world away!*

How?

Edward Lorenz, a meteorologist made a profound discovery in 1961.

He illustrated this effect with the analogy of a butterfly flapping its wings and thereby causing the formation of a hurricane half a world away.

A nice way to see this *butterfly effect* for yourself is with a game of billiards. No matter how consistent you are with the first shot (the break), the smallest of differences in the speed and angle with which you strike the white ball will cause the pack of billiard balls to scatter in wildly different directions every time. The smallest of differences are producing large effects - the hallmark of the butterfly effect.

To understand this phenomenon better, *Edward Lorenz*, a meteorologist conducted experiments in weather forecasting from past data on his computer. He created a mathematical model which, when supplied with a set of numbers representing the current weather, could predict the weather *a few minutes in advance.*

Once this computer programme was up and running, Lorenz could produce long term forecasts by feeding the predicted weather back into the computer *over and over again* with each run forecasting further into the future. Accurate minute by minute forecasts added up into days, and then weeks. One day, Lorenz decided to rerun one of his forecasts. To save time he decided *not* to start from the scratch; instead he took the computer's prediction from the middle point of the first run as the start. He was amazed at his findings:

*Although the computer's new predictions were initially the same as before, the two sets of predictions soon began to change dramatically!*

Why?

Lorenz realized that while the computer was printing the predictions to three decimal places, it was actually *crunching the numbers internally to six decimal places.*

So although Lorenz had started the second run with the number 0.415, the original run had used 0.415129.

This seemingly insignificant difference corresponds to the sort of difference that the flap of a butterfly's wing might make to the breeze on your face.

The conclusion is that though the starting weather conditions in the two cited instances are *virtually* identical, the predictions were different, the quantum of difference depending on *how long* the experiment is conducted. Small errors in the measurement of the current weather would *not* remain small but increase relentlessly in size each time they were fed back into the computer.

Lorenz illustrated this effect with the analogy of a butterfly flapping its wings causing the formation of a hurricane half a world away. In this context, the reader is reminded of the billiards analogy we mentioned earlier.

*No matter how consistent a player is, the smallest difference in the first strike of the white ball will cause the other balls to scatter in wildly different directions every time.*

This is a good example of a *chaotic system*. It is worth noting that the laws of physics that determine how the billiard balls move are precise and unambiguous with no possibility for *randomness*. What at first glance appears to be random behaviour is completely *deterministic* - it only seems random because imperceptible tiny changes are causing significant variations over time. The rate these tiny differences add up provides the chaotic system with a *prediction horizon* - a length of time beyond which we can no longer accurately forecast its behaviour.

In the case of the weather, the prediction horizon is about one week. It seems this is the ultimate limit for computers and software.

(50 years back it was less than 24 hours.

Surprisingly, our parent solar system, with a prediction horizon of a *hundred million years* is probably the first chaotic system to be discovered)

French mathematician *Henri Poincare* in 1887 postulated that while Newton's Theory of Gravity could predict how two planetary bodies would orbit under their mutual attraction, adding a third body to the mix rendered the equations *virtually unsolvable*. The best we can do for three bodies is to predict their movements moment by moment, and feed the data back into our equations…

Though the dance of the planets has a lengthy prediction horizon, the effects of chaos cannot be ignored e.g. in the trajectories of *asteroids*. Keeping an eye on asteroids is difficult but necessary for our survival considering that an asteroid strike had wiped out dinosaurs from the face of the earth 65 million years ago.

With modern technology we can safeguard against cosmic hazards like *steering comets away from a potential collision with Earth*.

Stability is required in many scenarios such as flying. Commercial aircraft are aerodynamically stable, so that a small turbulent nudge (butterfly effect) will not push the plane out of a level flight path.

Fighter pilots find this stability to be a hindrance who like their aircraft to make rapid changes with minimal effort.

Modern fighter aircraft are equipped with on board computers that constantly adjust the flight surfaces to cancel unwanted butterfly effects giving the pilot greater flexibility to maneuver his aircraft. The key to unlocking the hidden structure of a chaotic system is in determining its preferred set of behaviour - known to mathematicians as its *attractor*. The mathematician *Ian Stewart* explains the phenomenon further:

Imagine taking a ping pong ball to the ocean and letting it go. If released above water it will fall and float if released below. In both cases, the ball will move in a *predictable* manner i.e. quickly to its attractor, the *ocean surface*. On the surface it will move in a *chaotic* manner.

However it is not *total* chaos. For example we do know how the ping pong ball will behave if jolted off its attractor.

Mathematicians use the expression *phase space* to describe this movement in geometric terms. Phase space is not *always* like regular space - each location in phase space corresponds to a different configuration of the system.

In phase space, a stable system will move predictably towards a *simple attractor* that is either a single point if the system settles down or a simple loop if the system cycles between different configurations repeatedly. A chaotic system will also move predictably towards its attractor in phase space - but instead of points or simple loops, we see *strange attractors* – complex and beautiful shapes (called *fractals*) that twist and turn in intricate detail.

*Fractal mathematics* was pioneered by French American mathematician *Benoit Mandelbrot*. It allows us to come to grips with the preferred behaviour of a system even as the incredibly intricate shape of the attractor prevents us from predicting *exactly* how the system will evolve once it reaches.

Is phase space abstract and unrelated to our day to day lives?

*Not at all.*

As an example, it helps to understand how our heart beats. The millions of cells that make up our heart are constantly contracting and relaxing separately as part of an intricate chaotic system with *complicated attractors*. These millions of cells must work in sync, contracting in just the right sequence at just the right time to produce a *healthy heartbeat*. Fortunately, this intricate state of synchronization is an attractor of the system - but it is not the only one. If the system is jolted somehow, it may find itself on to an altogether different attractor called *fibrillation*, in which the cells constantly contract and relax in the wrong sequence.

The purpose of a *defibrillator* is to apply a large voltage of electricity across the heart. This is not to *restart* the heart cells as such but to give the chaotic system enough *kick* to move it off the fibrillating attractor, back to the healthy heartbeat attractor. The main benefit of having a chaotic heart is that tiny variations in the way those millions of cells contract serves to distribute the load more evenly, reducing wear and tear on the heart and allowing it to pump decades longer than would otherwise be possible.

*Chaos Theory* is not the sole preserve of mathematicians. It draws together specialists from diverse fields like physicists, biologists, computer scientists and economists. Not only can chaotic systems be found almost anywhere, they share many common features independently of where they come from.

Consider both a dripping tap and the *supercooled liquid helium* used as a coolant in the Large Hadron Collider. Both are non chaotic systems at the start, but as you slowly heat the helium, tiny convection cells begin to form; as you slowly open the tap the character of the dripping sounds will change.

Eventually the increases in temperature and water flow will cascade into the *chaos of boiling helium and rushing water.*

The transition from *order* to *chaos* in these systems is controlled by an exact number- the *Feigenbaum Constant*.

From dripping taps to the Large Hadron Collider, from a beating heart to the dance of the planets, chaos is *all around us*. Chaos Theory has turned our attention to things we once thought we understood, and shown us that *nature is far more complex and surprising than we had ever imagined.*

# Twenty Five

# Pi(∏), Phi(Φ) And The Golden Ratio

*The elegance of a mathematical theorem is directly proportional to the number of independent ideas one can see in the theorem and inversely proportional to the effort it takes to see them.*

**George Polya (1887 – 1985)**
**Hungarian mathematician specializing in number**
**and probability theories.**

The numbers with unusual properties are listed below:

The commonplace but infinite, transcendental number *Pi* ($\pi$)

The rare and mysterious number *Phi* ($\varphi$)

The Golden Ratio or *Fibonacci Sequence.*

*The all too familiar number Pi* ($\pi$):

March 14 is celebrated the world over as *Pi day* because 3.14 are the first 3 digits of Pi the number depicting the ratio of the circumference of a circle to its diameter.

The number is unique in that the digits following the decimal point can continue endlessly *without repeating itself!*

The current world record in deciphering Pi was set up by a personal computer in Japan that has calculated the *first five trillion* digits.

There are other fractions that are endless like 1/3 and 1/7 that do not attract so much attention.

Why?

1/3 is 0.3333333..... and is boring.

1/7 is 0.142857 looped over and over again making its predictability boring.

1/3 and 1/7 can be classified like any other fraction as *rational* numbers.

Pi is classified as an *irrational* number because the digits will *never* slip into a predictable pattern. Even if you have five trillion digits in front of you, you can't guess what the next digit will be without starting from the scratch.

Pi is considered *transcendental* since you could have a Pi day birthday party confident that somewhere down the line a string of numbers would appear representing *your birthday.*

The transcendental combination of an aperiodic string of numbers that extends for infinity means that this entire essay as numbers (a=1, b=2.......y=25, z=26) has already existed in Pi since the beginning of the universe.

Everything you do in your life is already written out in vivid detail in Pi.

The innocuous looking Pi is much more than what you think it is.

*The rather unfamiliar Phi (φ):*

Unlike its popular cousin Pi, not many would have heard of *Phi.*

Phi is also known as the *Golden Ratio.*

Like Pi, Phi is an irrational number meaning that its terms go on forever after the decimal point *without repeating.*

Phi can be defined by breaking a stick into two unequal pieces. If the ratio of the lengths of the 2 pieces is *equal* to the ratio of the whole stick to the longer piece, the ratio is termed the *Golden Ratio.*

This drab definition does not bring out the *uniqueness and beauty* of the Golden Ratio. Let us probe further:

The story of the GR is legendary. With a history dating back almost to the time of Pi, scholars including Pythagoras and Euclid have called it the golden ratio or *divine section*

The ratio is based on the relationship between consecutive numbers in a *Fibonacci Sequence.*

Each number in the sequence is the *sum of the two numbers before it* as follows:1, 1, 2, 3, 5, 8, 13, 21, 34, 55…. and so on *endlessly.*

$\varphi$ has been known since the time of the ancient Greeks. It was first described by the mathematician Euclid who called it *the division in extreme and mean ratio.*

You can think of $\varphi$ as a number that can be squared by adding one to the number i.e.

$$\Phi^2 = \varphi + 1$$

Expressed as a quadratic equation it leads to two solutions:

(1+ √ of 5) / 2 and

(1 –√ of 5) / 2

The first solution yields the positive irrational number *plus* 1.6180339887…..

The negative solution is *minus*

1.6180339887….

An elegant way to represent $\varphi$ is

$5^{1/2}$ x 0.5 + 0.5 or $5^{1/2}$ x 0.5 - 0.5

Reverting to the interesting relationship with the *Fibonacci* Sequence, it is easy to observe that:

The ratio of successive Fibonacci numbers gets *closer and closer* to $\varphi$ *plus* 0.6180339.

Interestingly, if you go in the reverse direction you get closer and closer to the *negative* solution i.e. $\varphi$ *minus* 0.6180339887….

Italian mathematician Luca Pacioli wrote a book called *De Divina Proportione* in 1509 on $\varphi$. He used drawings by Leonardo da Vinci who is believed to have used the expression *sectio aurea* (golden section) to describe them.

Three centuries later, American mathematician, Mark Barr used the Greek letter $\varphi$ to represent this number.

φ in diagrams:

Figure 1

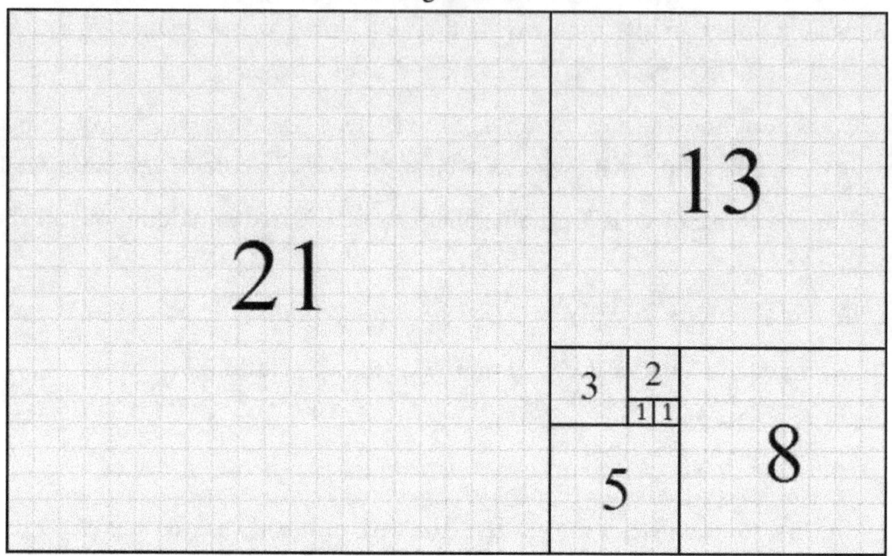


The Figure above is a graphic representation that can be repeated *infinitely*. The numbers in the squares are the Fibonacci numbers defined as 1,1,2,3,5,8,13,2……. to infinity. Each number is derived by adding the preceding two numbers e.g. 2=1+1, 3=1+2, 5=2+3 and so on as shown in Figure 1. A graphical representation of the numbers constitutes a *spiral* as can be seen in Figure 2 below:

It is the effect created by joining *opposite corners* of the squares. You get a *logarithmic spiral*.

In view of its intrinsic beauty and that it can be expanded *infinitely* it is also called the *Golden Spiral*.

φ in Nature:

Math enthusiasts describe the Fibonacci Sequence as *Mother Nature's Favourite Number Sequence*. The spiral of seeds in a pine cone, the fruitlets of a pineapple conform to the FS.

The sequence can be described by the equation:

$F(n) = F(n-1) + F(n-2)$ where $n > 1$ so,

$F(0) = 0$, $F(1) = 1$ and $F(2) = F(1) + F(0) = 1$

The sequence of numbers comprising the FS is:

0, 1, 1, 2, 3, 5, 8, 13, 21, 34, 55, 89, 144, 233, 377, 610, 987 ......

At first sight this may appear as a meaningless jumble of increasing numerals, but in real they are the first few numbers of the fascinating FS.

The person who brought the FS to Western audiences is *Leonardo of Pisa* who was born around 1170 A.D. and died around 1250 A.D. He was later nicknamed Fibonacci from *Filius Bonacci* meaning *son of Bonacci*.

What is not generally known is that the sequence had actually been derived by *Indian and Arab mathematicians ten centuries earlier.*

In 1202, Fibonacci described the sequence in his *Liber Abaci* (Book of Calculation) intended as a math guide for tradesmen for profit and loss calculations.

Reverting to the FS in nature, some flowers have 3, 5, 8, or 13 petals where each petal gets maximum exposure to sunlight. The rows of seeds in sunflowers and pinecones often add up to Fibonacci Numbers, because that is *the most efficient way to pack the maximum number of seeds into a small space.*

Figure 3

A picture of the spiral packing of Sunflower seeds.

Source: *Anna Benczur/ Wikimedia Commons*

The *Golden Ratio* can perhaps even be seen in the *Great Pyramid of Giza* (built between 2589 and 2504 BC) where the length of each side of the pyramid's base is 756 feet, and its height is 481 feet. The ratio of the base to the height is roughly 1.5717 that is close to the Golden Ratio.

Ancient Greek sculptor *Phidias* (500 BC - 432 BC) is said to have applied Phi to the design of the sculptures he created for the *Parthenon*.

*Plato* (428 BC - 347 BC) celebrated the Golden Ratio and *Euclid* (365 BC - 300 BC) linked it to the construction of a five sided figure the *pentagram*.

In the modern era of science (1970s) British physicist and mathematician *Roger Penrose* included the Golden Ratio in his *Golden Tiles* which allowed surfaces to be tiled in *five fold symmetry*. In the 1980s $\varphi$ was theorized to have appeared in *quasicrystals* a then newly discovered form of matter.

Studies have shown that when test subjects view a series of faces, the ones they deem the most attractive have Golden Ratio proportions between the width of the face and the width of the eyes, nose, and eyebrows.

Aside from the above instances, the Golden Ratio is found most frequently in plants to enable them to maximize the exposure of their leaves to the sun. This requires them to grow at *non repeating angles*. The easiest way to guarantee this is to have an irrational value for the number of leaves, and many of the spirals we see in nature are a consequence of this behaviour i.e. their distributions follow *logarithmic spirals*, the general mathematical form of a golden spiral.

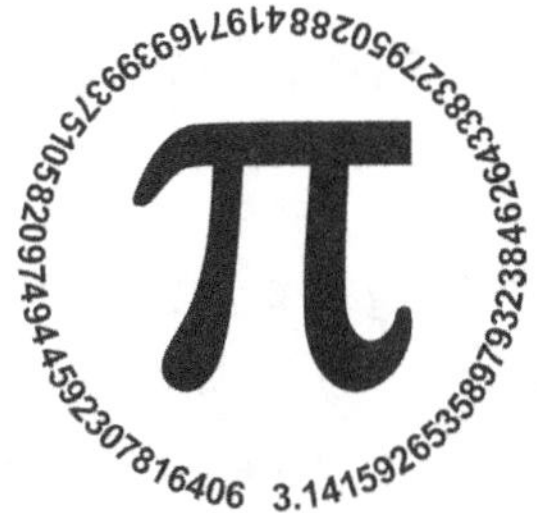

**Illustration 15:** The Symbol Pi (π) - Shutterstock _362948072

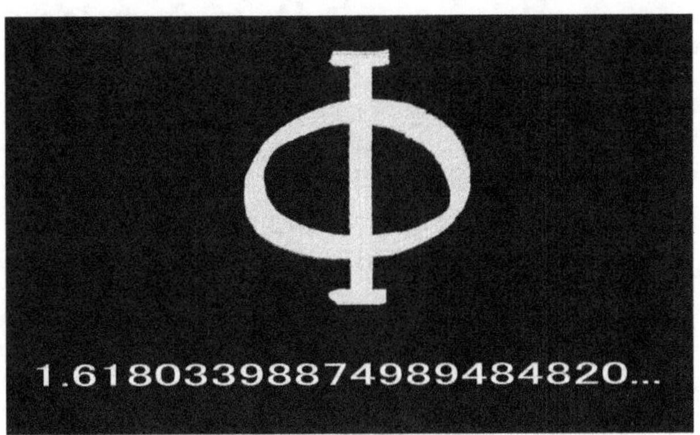


**Illustration 16:** The Symbol Phi (φ)- Shutterstock _2146943431

# Twenty Six

# The Ramanujan Magic Square

*The peculiar interest of magic squares in general lies in the fact that they possess the charm of mystery. They appear to betray some hidden intelligence which by a preconceived plan produces the impression of intentional design, a phenomenon which finds its close analogue in nature.*

**Paul Carus (1852 -1919)**
**American author and philosopher**

There are many magic square that Quiz Masters use in their shows. The square given below is the square attributed to none other than the Indian math genius Srinivas Ramanujan (1887 – 1920)

Figure 1

RAMANUJAN'S MAGIC SQUARE

| 22 | 12 | 18 | 87 |
| 88 | 17 | 9 | 25 |
| 10 | 24 | 89 | 16 |
| 19 | 86 | 23 | 11 |

This square looks like any other normal magic square. But this is formed by great mathematician of our country – Srinivasa Ramanujan.

What is so great in it?

First let us check whether it satisfies the matrix/trace rule we defined earlier:

Trace 1- Left to diagonal,

22 + 17 + 89 + 11 = 139

Trace 2- Right to left diagonal,

87 + 9 + 24 + 19 = 139

Please note that in both cases the trace is a *prime number.*

So if you are conducting a quiz show with the Ramanujan Magic Square the number you would pick out of the hat would be the prime number 139. We move on to the next 2 versions of the

RMS where the shaded portions would also add up to the prime number 139.

First let us check whether it satisfies the matrix/trace rule we defined earlier:

Trace 1- Left to diagonal,

22 + 17 + 89 + 11 = 139

Figure 2

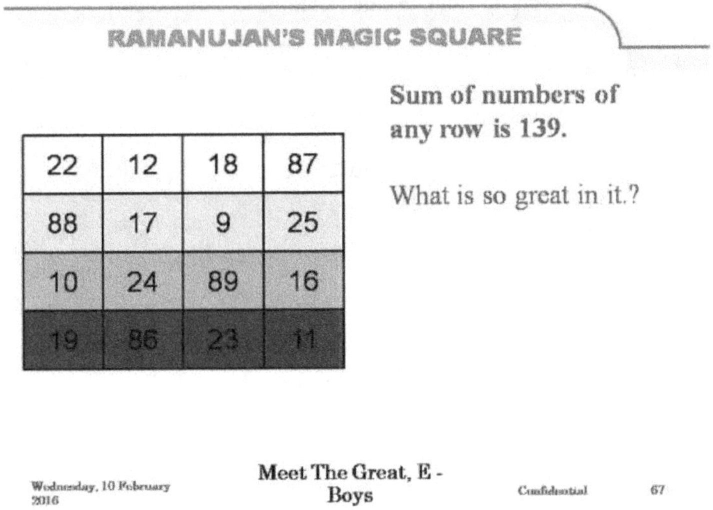

RAMANUJAN'S MAGIC SQUARE

| 22 | 12 | 18 | 87 |
| --- | --- | --- | --- |
| 88 | 17 | 9 | 25 |
| 10 | 24 | 89 | 16 |
| 19 | 86 | 23 | 11 |

Sum of numbers of any row is 139.

What is so great in it.?

Wednesday, 10 February 2016    **Meet The Great, E - Boys**    Confidential    67

Trace 2- Right to left diagonal

87 + 9 + 24 + 19 = 139

Please note that in both cases the trace is a *prime number.*

So if you are conducting a quiz show with the Ramanujan Magic Square the number you would pick out of the hat would be the prime number 139. We move on to the next 2 versions of the RMS (Figures 2 &3) that *any row or any column also adds up to 139.)*

Figure 3

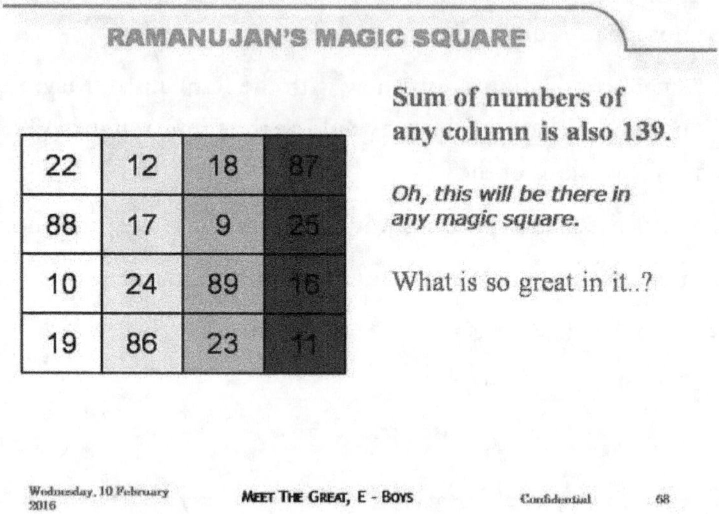

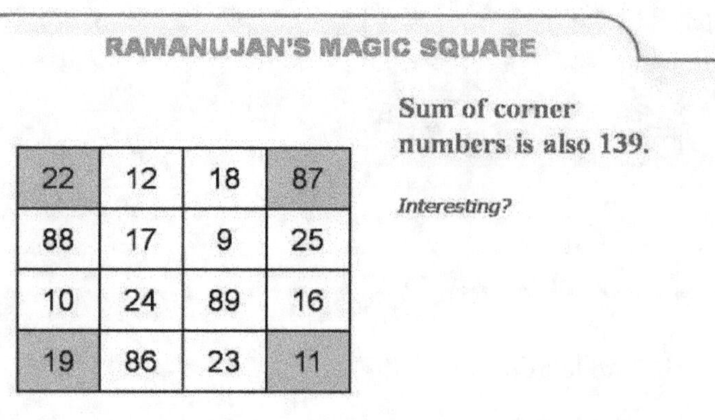



Similarly, in Figures 4,5,6,7, 8 and 9 appearing below the shaded portions indicate other box combinations that add up to 139:

Figure 4

Figure 5

## RAMANUJAN'S MAGIC SQUARE

| 22 | 12 | 18 | 87 |
|----|----|----|----|
| 88 | 17 | 9  | 25 |
| 10 | 24 | 89 | 16 |
| 19 | 86 | 23 | 11 |

Look at these possibilities. Sum of identical coloured boxes is also 139.

*Interesting..?*



Figure 6

## RAMANUJAN'S MAGIC SQUARE

| 22 | 12 | 18 | 87 |
|----|----|----|----|
| 88 | 17 | 9  | 25 |
| 10 | 24 | 89 | 16 |
| 19 | 86 | 23 | 11 |

Look at these central squares.

*Interesting...?*

Figure 7

## RAMANUJAN'S MAGIC SQUARE

| 22 | 12 | 18 | 87 |
|----|----|----|----|
| 88 | 17 | 9  | 25 |
| 10 | 24 | 89 | 16 |
| 19 | 86 | 23 | 11 |

Can you try these combinations?

*Interesting.....?*



Figure 8

## RAMANUJAN'S MAGIC SQUARE

| 22 | 12 | 18 | 87 |
|----|----|----|----|
| 88 | 17 | 9  | 25 |
| 10 | 24 | 89 | 16 |
| 19 | 86 | 23 | 11 |

Look at these possibilities. Sum of identical coloured boxes is also 139.

*Interesting..?*

## RAMANUJAN'S MAGIC SQUARE

| 22 | 12 | 18 | 87 |
|----|----|----|----|
| 88 | 17 | 9  | 25 |
| 10 | 24 | 89 | 16 |
| 19 | 86 | 23 | 11 |

**Can you try these combinations?**

*Interesting.....?*

Wednesday, 10 February 2016        MEET THE GREAT, E - BOYS        Meet The Great 74

Figure 9

We have now done a full circle and come back to the RMS with which we started. The square is special because the first row 22 12 18 87 *is the birthday of Srinivas Ramanujan.*

I shall now show you how to devise a magic square with the the first row of the 4x4 square showing *your birthday.* It need not add up to a prime number, but if it does you are in very *select group* having the likes of Srinivas Ramanujan.

Let us assume your birthday is 1ˢᵗ March 1948. The first row of the 4x4 square would be:

1 3 19 48

To fill the remaining rows we shall use the following nomenclature:

DD stands for *day*, 1 in the first row above,

MM stands for *month*, 3 in the first row above,

CC stands for *century*, 19 in the first row above and

YY stands for *year*, 48 in the first row above.

Formula for row 2:

DD is YY + 1 = 49

MM is CC - 1 = 18

CC is MM - 3 = 0

YY is MM + 1 = 4

The second row is 49 18 0 4

The third row is 1 3 50 17

( MM -2, MM, YY + 2 and CC - 2 )

The fourth row is 20 47 2 2

( CC + 1, YY - 1, MM -2 and MM -2 )

The completed square is:

1 3 19 48

49 18 0 4

1 3 50 17

20 47 2 2

In the above matrix, trace 1(left to right diagonal) and trace 2(right to left diagonal) add up to the *prime number* 71.

The person with the birthday 1ˢᵗ March 1948 is in the select group of Srinivas Ramanujan.

If you check all the other features of the RMS like sum of each row, each column, smaller designated squares within the main 4x4 square *equally hold good!*

The formula for the *magic square* that bears his name was discovered by none other than Srinivas Ramanujan himself.

Why not evolve a magic square with the first row depicting *your* birthday?

# Twenty Seven

# The Problem Of The Grazing Goat

*Only those who attempt the absurd are capable of achieving the impossible.*

**Miguel de Cervantes (1547 – 1616)**

Miguel de Cervantes is the Spanish writer best known for his novel *Don Quixote.*

This is a centuries old problem that has been partially resolved recently. The problem itself seems deceptively simple.

It is described through an intriguing thought experiment:

Imagine a circular fence that encloses one acre of grass. If you tether a goat to a point on the *inside* of the fence, how long a rope would you need to allow the animal access to *exactly* half an acre of grass?

Why is the chosen animal a goat?

A bull or a sheep, with a shorter neck would have been a more appropriate choice. Unlike the goat they prefer to *graze* than *browse*, or put in another way chew food from the ground than from a height. Be that as it may, mathematicians for the past three centuries have persisted with the ubiquitous goat in analyzing this problem and that practice has stuck.

The problem as stated seems right out of a high school geometry text book and yet.

*Math enthusiasts have been grappling for a solution for the past 270 years. They have succeeded in solving some versions. However the bottom line is that the goat in a circular fence problem has refused to yield anything but fuzzy, incomplete answers.*

For a better understanding of the problem, we need to probe further on related historical developments in the 21st century:

In 2017, *Ingo Ullisch*, a mathematician from the University of Munster, in North Rhine - Westphalia, Germany claimed he had found an exact solution.

It was known by then that the 2 dimensional goat problem could be reduced to a single dimension on paper using a *transcendental* equation (an equation with polynomial, trigonometric, exponential, calculus like mathematical functions)

This created a new set of problems since some equations that arose like x = cos(x) had no *exact* solution. Ullisch circumvented this hurdle by working with the more *tractable* transcendental equation $\sin(x) - x \cos(x) - \pi/2 = 0$.

He used *complex analysis*, an available mathematical tool applying calculus to expressions containing complex numbers. In a lighter vein he remarked - *I am the first mathematician to apply complex analysis to hungry goats.*

With this strategy he was able to transform his transcendental equation to an equivalent expression for the length of rope that would let the goat graze in half the enclosure.

The reviews were mixed.

*Michael Harrison,* a mathematician at Carnegie Mellon University described it as an *advance* in the solution of the grazing goat problem but nothing more.

The reason for this *conditional* acceptance is that there is a *catch* in the solution provided by Ullisch. Delving further in this line of thinking will suck us deeper into abstract mathematics which is likely to scare the lay reader of this essay. To retain reader interest I am taking the liberty of disclosing the solution:

*The rope should be approximately 1.16 times the radius of the field to allow the goat to graze half the field.*

(As explained earlier the field is circular measuring one acre, and the goat is tethered on the inside of the fence)

It is evident that the solution is approximate, *not* exact.

I conclude this essay with an apology to the reader for missing out on the mathematics jugglery that led Ingo Ullisch to the solution. He concedes its limitations in the following words:

*This problem is an isolated one. It is not connected to other problems to be generalized within a mathematical theory.*

The positive aspect is that even fun puzzles like this give rise to new mathematical ideas and help researchers attempt innovative approaches to find solutions to fresh problems.

**Illustration 17:** The Problem of the Grazing Goat- Shutterstock _1025159959

# Twenty Eight

# The Hare & The Tortoise And Infinity Puzzles

Mathematics is not a popular subject at the school level.

To popularise it, new techniques have been developed by specialists in Europe and America. Such techniques are expected to be increasingly used. Franchises are being awarded to education related entrepreneurs in many developing countries across the globe.

There are reports of increasing interest and successful sales promotion of the new techniques.

The reason for the dislike among children for math is not far to seek.

When learning history, the child can relate easily to the rule by kings, queens, emperors a few centuries back. In the modern age they learn of presidents, prime ministers, democracies and dictatorships.

In all cases, there is a personal touch that appeals to the growing child.

Geography is interesting because it enlightens the child on the location of and physical features of different countries such as mountains, lakes, rivers, deserts which can be seen, felt or imagined. The natural beauty of each location besides its seasonal variations are easy to relate to. The same can be said, though to a lesser extent of subjects like economics, politics, journalism, engineering, medical sciences and affiliated subjects.

In contrast, mathematics is totally 'impersonal', unconnected directly to the worldly objects around us, past history, present day events and future occurrences.

It is based on a logic of its own using a plethora of Greek symbols, formulae, and equations to which the child is introduced at school.

At the college level it becomes more abstract branching off to the study of calculus, vectors, and tensors and many other abstract entities.

A question often asked is

'Why do we need to study math in the first place. Is not arithmetic (addition, subtraction, multiplication, and division) enough to conduct our day to day lives'?

The answer is yes and no.

Yes - because day to day living does not require higher math per se!

No - because you cannot master physics or even chemistry without a knowledge of higher math!

Math is essential to 'decipher' and 'predict' a result in modern physics and related sciences.

Trying to master modern physics without math is like 'climbing a roof without a ladder'

Let me explain further:

Math is by nature, an abstract subject, and it can be hard to wrap your head around it if you don't have a good teacher to guide you. Be that as it may, few would deny its role as a vital factor in the rapid evolution of society. We reached the moon because of math, learnt the secrets of DNA, transmitted electricity over thousands of miles, besides creating computers. These achievements would have been impossible without math. Today we see its effects reverberating in cellphones, satellites, hula hoops and automobiles.

We give below the names of 5 of the greatest mathematicians of all time and their principal contributions:

1 Isaac Newton (1642 - 1727) - CALCULUS

2 Carl Gauss (1777 - 1855) - NUMBER THEORY & COMPUTERS

3 John Von Neumann (1903 - 1957) - ARCHITECT OF THE MODERN COMPUTER

4 Allan Turing (1912 - 1954) - CRYPTOLOGY IN COMPUTER SCIENCE

5 Benoit Mandelbrot (1924 - 2010) - FRACTAL GEOMETRY

The significance of the above contributions is not self explanatory for the layman! Details will come only later in our narrative when the context is appropriate. Meanwhile, I give below a few interesting aspects of math, the cases increasing in complexity that I hope will reveal its beauty and hidden message to the world:

## I THE ENGLISH ALPHABET AND ITS NUMERICAL SIGNIFICANCE:

We nominate numbers for each alphabet in the order in which they appear, A = 1, B = 2, C = 3.....................X = 24, Y= 25, Z = 26 You can create a Ready Reckoner covering all 26 alphabets for reference purposes. We next evaluate words we frequently use in our day to day lives. I have chosen the following 6 words:

L U C K 12+21+3+11 = 47%

L O V E 12+15+22+5 = 54%

M O N E Y 13+15+14+5+25 = 72%

KNOWLEDGE 11+14+15+23+12+5+4+7+5 = 96%

H A R D W O R K 8+1+18+4+23+15 +18+11 = 98%

A T T I T U D E 1+20+20+9+20+21+4+5 = 100%

The example cited above may appear specious to some but on reflection it is generally accepted that having the right "ATTITUDE" towards the problems we face in life is the key to success and happiness. This is supported by the alphabets in the English language and the order in which they appear I.e. A = 1.............. Z = 26

Yet another unusual feature of the connection of numbers to the 26 alphabets in the English language: In order to cover usage of the first 4 alphabets a, b, c, and d we need to count up to 1 "billion". In the range 1 - 100 'd' occurs first in 'hundred'

In the range 1 - 1000, the letter 'a' appears first in 'thousand'

In the range 1 - 1,00,000,000 (1crore) 'c' appears first in 'crore'.

In the range 1 - 1000, 000,000) the letter 'b' appears first in 'billion'

Is this not amazing?

## II ACHILLES (THE HARE) AND THE TORTOISE:

Consider a race between sprinter Achilles (The Hare) and the proverbially slow tortoise. The tortoise enjoys a 100 metre advantage when the race commences. As per the logic of modern math Achilles (The Hare) cannot overtake the tortoise even after several hours!! Why? In order to overtake the tortoise, Achilles (TH) needs to cover in the first instance the 100 meter gap separating them. Achilles (TH) will probably do it in 12 seconds. During this 12 seconds the tortoise has moved forward by say 1 metre. Achilles (TH) needs to cover this 1 metre to overtake the tortoise. A little thought will show that this line of reasoning is 'endless' because the tortoise always enjoys the 'right to move first'

We know from commonsense that Achilles (TH) will overtake the tortoise soon after 12 seconds has elapsed.

The explanation lies in the 'infinite reducing gaps' between Achilles (TH) and the tortoise. This constitutes an "infinite" reducing geometric series with a "finite" sum.

Once the finite sum is crossed Achilles (The Hare) will moved ahead of the tortoise!!

## III HANDS OF A CLOCK

The question is at what time between 7 and 8 are the hour hand and minute hand in a perfect straight line but not together. When will a 'perfect diameter' be formed between 7 and 8 on the circular face of the clock? The spontaneous answer would be at 5 minutes past seven. However as in the case of Achilles and the tortoise, the slower hour hand always has the "right to move first".

The ACHILLES / TORTOISE and HANDS OF A CLOCK are instances of an "infinite " geometric sequence having a "finite" sum. Reverting to The Hands of a Clock example, the "perfect diameter" is formed not at 7:05 but at 7:05:27 i.e. 5 minutes and 27 seconds past 7!

In math aside from geometric sequences there are "arithmetic", "harmonic", and "Fibonacci" sequences. We shall briefly describe them with examples:

III Arithmetic Sequence:

Have you during schooling, considered what would be the sum of the first hundred numbers, 1+2+3...........+98+99+100?

You can of course laboriously add the numbers in the order that they appear and arrive at the result 5050.

There is a simpler way:

Suppose you add the first number and the last,

1+100 = 101 followed by the second and second last and so on till you reach 50+51.

In each case you will get the same result 101, and since there are 50 combinations the total would be 101x50 or 5050.

The above example shows a typical "arithmetic sequence" Please note that each item in the sequence is obtained by "adding" a factor to the preceding number.

(In the geometric sequence each item is obtained by "multiplying" the preceding number by a factor.)

There are many remarkable characteristics to the arithmetic sequence that we have chosen:

The sum of 100 numbers is 5050

The sum of 1000 numbers is 500500

The sum of 10000 numbers is 50005000

The sum of 100000 numbers is 5000050000!

Harmonic Sequence: This is an arithmetic sequence in reciprocal form like,

1/10, 1/20, 1/30, 1/40.........

An example of a harmonic sequence in day to day life is in the vibrating string of the violin. The natural harmonic is the pitch (resultant) that is produced by lightly touching an open vibrating string (the fundamental) at one of the nodes located at 1/2, 1/3, 1/4, etc the length of the string.

Fibonacci Sequence: These numbers form a sequence where any number is the sum of the preceding two numbers e.g.

1,1,2,3,5,8,13,21,34,55........

What is not generally not known is that examples of the Fibonacci Sequence can be seen all around us like in the number of spirals in a pine cone or pine apple, seeds in a sunflower or petals in flowers:

Most flowers have three (lilies and irises), five (parnassia, rose hips) or eight (cosmea), 13 (some daisies), 21 (chicory), 34, 55 or 89 (asteraceae).

In a graph the Fibonacci numbers show a "spiral pattern" noticed in HURRICANE RITA as it approached the Louisiana and Texas shore on Sept 23 2005.

Mathematicians refer to the Fibonacci graph pattern as the GOLDEN SPIRAL.

The Fibonacci Sequence is mentioned in the year 1202 A.D. in the book LEBER ABACI by Italian mathematician Leonardo Fibonacci (1170 -1250) Surprisingly, the sequence is described about a millennium earlier in 200 B.C.E. Indian mathematics. The author is Acharya Pingala while enumerating possible patterns of Sanskrit poetry formed from syllables of two lengths!

## IV TRISECTION OF AN ANGLE:

At school we were taught how to bisect an angle using compass and ruler 'only' The bisected angle can itself be bisected i.e. quadrasection is possible which goes on and on.

What is generally not known is that using compass and ruler only it is IMPOSSIBLE to trisect an angle!!

## V THE MISSING RUPEE:

Three friends go to a restaurant for a coffee. They get a bill for Rs.30/- which is paid and the trio leave the restaurant.

As a gesture, the owner returns Rs5/- to the trio through his assistant.

The assistant returns Rs 3/- to the trio and pockets the Rs 2/-

Now comes the catch.

The three friends have initially paid Rs 10/- and got back Rs 1/- each i.e. they have paid Rs 9/- each totalling Rs 27/-

The only other money in circulation is Rs2/- that the assistant pocketed.

Where is the missing rupee?

## VI THEOREMS AND RIDERS:

At school in Geometry Class we learnt 'therems' i.e. geometric truths.

A good example is the Theorem of Pythagoras that states that the square of the hypotenuse of a right angled triangle is equal to the sum of the squares of the other two sides.

A Rider is a problem based on a particular theorem. It cannot be solved unless the theorem on which It is based is fully understood.

A favourite rider of mine refers to a point anywhere within an equilateral triangle (a triangle where the three sides are equal and each of the three angles is 60 degrees.)

From any point within the triangle drop perpendiculars to each side. Also drop a perpendicular from any of the vertices to the opposite side.

The rider requires you to prove that the the sum of the three perpendiculars is equal to the perpendicular from the vertex to the opposite side.

The solution is deceptively simple and I leave it to the reader to figure it out.

I would like to end this essay with a quote by Silvanus Phillips (1851 -1916) professor of physics at the City and Guilds Technical College in Finsbury, England and author of the book CALCULUS MADE EASY published in the early 20th century. Thomson downplayed the difficulty in learning math with the introductory statement:

"What one fool can do another fool can also do"!!

This should go a long way in dispelling the unfounded fear of mathematics!!

**Illustration 18:** The Hare and the Tortoise- Shutterstock _1657400494

# Twenty Nine

# Million Dollar Prize For Resolving The P Vs Np Problem

A quality that distinguishes man from other life forms is his ability to think and analyze complex issues. It could be argued that lower life forms also think e.g. the complex architecture of a bird's nest or a beehive shows the ability to "plan" and "execute".

A little thought would clarify that these are products of instinct rather than intelligence!

Viewed from this angle, man is definitely endowed with special abilities as depicted by structures of excellence he has created like the Taj Mahal at Agra or the Empire State Building in N

ew York. They are products of elaborate planning and execution.

Mathematics is one such attribute that is special in the field of sciences being an "intangible"

Unlike its cousin physics it does not require observation of natural phenomena leading to theories that are verified through testing under varying conditions. The connection between physics and math is intimate and all pervading. Nobel Prize winner Eugene Wigner has this to say:

"Mathematics has the uncanny ability not only to describe and explain but to predict phenomena in the physical world.

Consider for a moment our day to day experiences in electricity (getting a shock) and magnetism (the kid's thrill of picking up iron filings with toy magnets)

The theoretical physicist is able to explain these and related phenomena by means of just four mathematical equations named after physicist James Clerk Maxwell.

Through these equations he showed in 1864 that varying electric or magnetic fields would generate waves in the electromagnetic spectrum that includes light, radio waves, X- rays,etc. These were eventually detected by Heinrich Hertz in the late 1880s.

At the astronomical macro level that applied equally to our day to day livings on planet earth, the Newtonian equations on the Laws of Motion and Gravitation of Classical Physics are well known. These were refined but not invalidated by Einstein's celebrated equation E = mc2 in the Special Theory of Relativity. The General Theory of Relativity described in mathematical terms by Einstein's Field Equations required a new brand of math termed "tensor calculus".

A remarkable example of the impact of math at the quantum level is the modern mathematical theory that describes how light and matter interact known as quantum electrodynamics (QED).

Mathematics is as old as the hills - going back centuries and millennia. Take the case of numbers that enable us to count. This depends on "zero",the "cardinal numbers" 1 to 9, and use of "decimals". Man has instinctively been attracted by the number 10 probably because of his ten fingers and toes!

Without such a system you would not be able to describe whole and fractional quantities of objects lying on the table before your very eyes.

The invention of zero is a subject of debate in scientific circles. It developed in India between the third and fifth century A.D. According to Peter Gobets, secretary of the ZERORIGINDIA Foundation based in the Netherlands,

"Though people have always understood the concept of nothing or having nothing, the concept of zero is relatively new; it fully developed in India early in the Christian Era. Before this mathematicians struggled to perform the simplest arithmetic calculations......

Today zero allows us to perform calculus, do complicated equations and work on computers. The numerical zero is widely seen as one of the greatest innovations in human history"

Apart from zero, mathematics/science is LOGIC, REASONING, deduction through observation, formulation of models with precise predictions, theorems, PROOFS, RESULTS, ABSOLUTE TRUTHS. It is a HARD SCIENCE.

There is of course the counter argument that humanities like philosophy also have the processes of LOGIC and REASONING. It is generally conceded though that within the sciences, mathematics is replete with absolute truths that are internally consistent. differs from other sciences in being entirely theoretical - a medley of Greek symbols. principles, rules used as tool to describe mainly the physical world but also in other disciplines like biology, chemistry, engineering and computers. As a tool as mentioned earlier it is so versatile to form the basis for physicists in to make predictions!

In Applied Mathematics, characteristics of the physical macro and micro worlds like the movement of celestial bodies at the macro level and micro extremes like the electron within the atom are made intelligible through a wide variety of equations, constants, and interpretations.

I have handpicked two interesting anecdotes/developments in the last two centuries in the field of mathematics:

I THE MAGIC NUMBER 1729

II THE MILLION DOLLAR PRIZE

The Magic Number 1729 is a legacy of the late Indian mathematician Srinivasa Ramanujam.

Born in the early 20th century to a poor family in the nondescript town of Erode in the southern part of the then British India, Ramanujam had a difficult childhood.

At school he was considered to be quite ordinary as he took no interest in any subject other than mathematics.

In math he was way ahead of his fellow students and even his teachers!

He intuited many complex end results that he did not know how to derive!

Poverty drove him to send a sample paper to G.H.Hardy, mathematician at Trinity College, Cambridge who recognized the genius in the young Indian.

Hardy organized a scholarship for Ramanujam to conduct research in mathematics at Cambridge and arranged his travel and stay in England.

During one of his visits to Ramanujam, Hardy had hired a cab with number plate ending 1729. There was a huge cultural gap between the two men; Hardy was at a loss what say.

Just to make conversation he casually remarked:

The number plate of my cab ends with the 'uninteresting' number 1729.

Ramanujam immediately shot back:

'On the contrary, it is unique- the smallest number that can be expressed as the sum of two cubes in two different ways'!

10 x 10 x 10 = 1000  9 x 9 x 9 = 729

TOTAL = 1729

12 x 12 x 12 = 1728  1 x 1 x 1 = 1

TOTAL = 1729

An example of Ramanujam's amazing insight on numbers!

As a matter of interest, the next three numbers sharing this property are:

4104, 13832, and 20683.

Can you figure out the cube duos that make up these three numbers?

II THE MILLION DOLLAR PRIZE

Despite the tremendous advancement in the sciences and math in the 21st century there are six nagging problems in math that have eluded solution till the present day. Matt Parker, Australian math enthusiast announced during one of his TV Shows that anyone who solved anyone of these problems would win a prize of 1 million USD from the Clay Institute of Mathematics, New Hampshire, USA!

At the beginning of the 21st century there were seven unresolved math problems. Grigori Perelman, a Russian mathematician solved THE POINCARE CONJECTURE and won the 1 million dollar prize in 2003.

The six unresolved problems are:

1.  THE BIRCH WINNERTON-DYER CONJECTURE

2.  THE HODGE CONJECTURE

3.  THE P VS NP PROBLEM

4.  RIEMANN HYPOTHESIS

5.  THE NAVIER-STOKES EXISTENCE AND SMOOTHNESS PROBLEM

6.  THE YANG-MILLS EXISTENCE AND MASS GAP PROBLEM

The problems are abstract and it is not possible to explain all of them.

I shall attempt to outline No.3 - P VS NP PROBLEM for the layperson:

P stands for 'Deterministic Polynomial Time'

NP stands for 'Non deterministic (Checking) Polynomial Time'

Solution to the P VS NP PROBLEM lies in proving conclusively that either

P is equal to NP or

P is not equal to NP

To understand the problem better let us take an example based on numbers (simplest polynomial):

If you ask a eight year old kid in school: 'What is 4 cubed? ' you are likely to get a 'quick' answer: 64(4 x 4 x4)- a P example

Instead if the same child is told:

'Pls check if there is a number which when cubed equals 64, the response may not be that 'quick' i.e. when compared to P!

NP may take longer than P i.e. there is a possibility that P is NOT equal to NP.

On the surface all this may sound like a silly anti climax with no possible connection with a 1 million USD prize!

Really?

Let us now SCALE UP the P/NP example. We shall examine the effect of 'increasing exponentials'

Suppose we started with a 20 digit number which is raised to the power 10. How would the first example play out?

Modern computing would flesh out both P and NP but would it take the same time?

We do not have a precise answer!

What is needed is an 'algorithm' that proves conclusively the answer to the query one way or the other!

Leading mathematicians of the world are in both camps; approx 100 proofs have been submitted. P not equal to NP enjoys a marginally higher number than P equals NP. None of the proofs have held up to scrutiny by experts.

Amongst the proofs submitted, Indian mathematician Vinay Deolalikar's submission (P is not equal to NP) created ripples of excitement with leading mathematicians the world over. After thorough scrutiny however the consensus is that in its present form Deolalikar's submission needs improvement on various fronts to make the grade.

A more recent development in this context is the claim made by Kumar Easwaran, a Hyderabad based mathematician of having found a proof for the Riemann Hypothesis that has remained unsolved for 161 years!

What is the RIEMANN HYPOTHESIS?

Without overwhelming readers with intricate math, RH delves around prime numbers. For the benefit of the layman, prime numbers cannot be expressed as the product of two other smaller numbers e.g. 2,3,5,7 are common examples.

RH in a nutshell gives a method of generating large random numbers and count prime numbers. The hypothesis originated from the research of the famous 19th century mathematician Carl Friedrich Gauss.

Easwaran a researcher at the Sreenidhi Institute of Science and Technology (SNIST) had made his claim in the Internet 5 years back. However editors of mathematical journals were unwilling to put his research out before the peer review.

A few years back his paper titled 'THE FINAL AND EXHAUSTIVE PROOF OF THE RIEMANN HYPOTHESIS FROM FIRST PRINCIPLES' was downloaded over a thousand times. A team of mathematicians called for a review of Easwaran's proof of the RH by 1200 peer mathematicians.

Interestingly only 7 of the 1200 mathematicians showed up and their opinion was that Easwaran's proof is correct. SNIST subsequently issued a statement stating that the RH and its proof will lead to proof of a number of other theorems that remain unproven as they are dependent on the proof of this particular hypothesis. It remains to be seen whether Easwaran will be awarded the $1 million prize!

The six problems (now possibly 5) are wide open and continue to pose a challenge to leading mathematicians!

# Thirty

# The Precocious Genius Of Friedrich Gauss

*It is not knowledge, but the act of learning, not possession but the act of getting there which grants the greatest enjoyment.*

**Carl Friedrich Gauss (1777 - 1855) German mathematician and physicist**

Gauss is considered one of the greatest mathematicians of all time. He was a humble self effacing person. When asked about his extraordinary contribution to mathematics, he is reported to have said:

*"The study of Euler's works will remain the best school for the different fields of mathematics and nothing else can replace it"*

(Leonhard Euler was a Swiss mathematician who died a few years after Gauss was born. He made significant contributions in *topology, analytical number theory* and *infinitesimal calculus*)

Be that as it may, Gauss was considered by the scientific community as the *Prince of Mathematics* and the *greatest mathematician since antiquity.*

Gauss corrected an error in his father's payroll calculations when he was just three years old!

At the age of seven he discovered the formula to calculate the sum of an arithmetic series that is used even today.

He studied at the prestigious University of Gottingen from 1795 to 1798. He was a pioneer in discovering hitherto unknown properties in *prime numbers* and *complex numbers* (combinations of real and imaginary numbers).

At the age of twenty two he proved the *Fundamental Theorem of Algebra* relating to complex numbers, closely followed two years later by his book *Disquisitiones Arithmeticae* considered today as one of the most influential mathematics books ever written. It laid the foundation for modern number theory.

Gauss showed an extraordinary interest in astronomy predicting the position of the *planetoid Ceres* that differed greatly from the view of contemporary astronomers. When Ceres was finally discovered in 1801 the Gauss prediction was proved to be accurate.

(At a personal level, in later years Gauss is stated to have become arrogant and unpleasant to the extent of claiming many discoveries of his juniors as his own.)

In the area of *probability* and statistics he introduced what is known today as the *Gaussian distribution, the Gaussian function* and the *Gaussian Error Curve.*

He made inroads into *modular arithmetic* used today in number theory, abstract algebra, cryptography and even in *visual and musical art.*

Some of his ideas clashed with Euclidean geometry that worked on the basis of a *flat* rather than on a *curved universe.* He did not like controversies and refrained from publicizing his ideas. Despite this he was considered by many as the father of non Euclidean geometry.

Apart from pure mathematics, Gauss invented the *heliotrope* used in land surveys and collaborated with *Wilhem Weber* to measure the Earth's magnetic field. Gauss passed away in 1855. Scrutiny of his unpublished papers revealed novel ideas that extended to the closing stages of the 19th century. Gauss as a scientific researcher was not a person who *yearned for recognition.*

**Illustration 19:** Carl Friedrich Gauss- Shutterstock _1516139771

# Thirty One

## The Seven Bridges Of Koningsberg

*E*ssentially Graph Theory, I have chosen this problem for top status.

*Is it possible to cross each of the seven bridges exactly once and come back to the starting point without swimming across the river?*

*The analysis that follows establishes Leonhard Euler as a mathematics matician of the highest order.*

I have not encountered a more severe brain teaser than the *The Seven Bridges of Koningsberg.*

The solution as explained by Leonhard Euler is based on mathematical principles and not a play on words or a conjuror's sleight of hand.

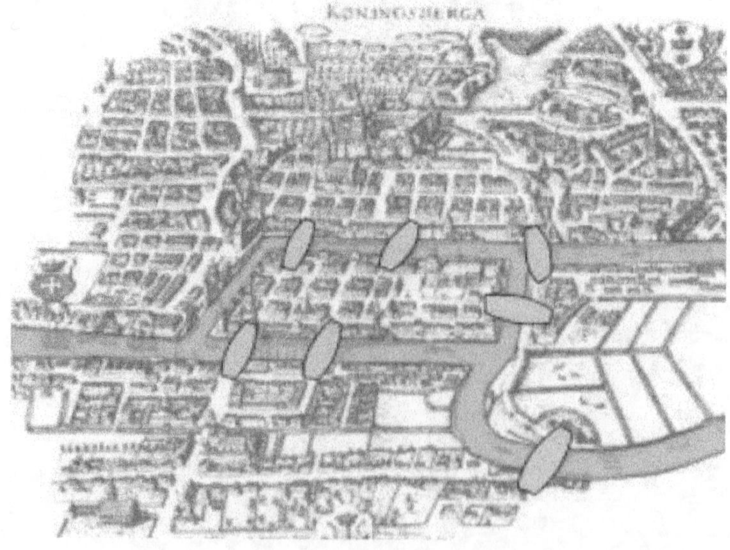

*The Koningsberg bridge problem asks if the seven bridges of the city of Koningsberg formerly in Germany but now known as Kalingrad in Russia over the Preger River can all be traversed in a single trip without doubling back?*

*(The trip has to end at the same point where it commenced)*

Figure 1…The ancient city of Koningsberg. The picture above gives an aerial view of the city, the River Preger, the seven bridges (green spots) and the habitable (shaded portions) areas of the city.

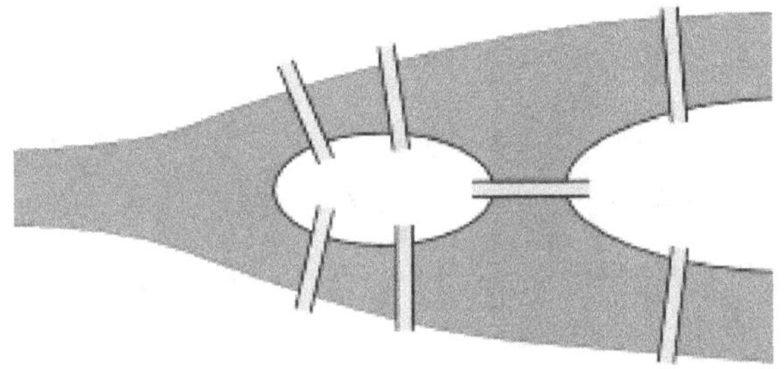

Figure 2

For a better view (Figure 2) outlines the four habitable areas of the city (white) north of the river, south of the river, on the island, and on the peninsula (on the right).

We will designate the areas as A, B, C and D respectively. (Figure 3)

Further the bridge will be crossed at points p, q, r, s, t, u, and v. (Figure 3)

To simplify further, instead of taking long walks through the town, you can visualize the scenario by drawing lines with a pencil (Figure 4)

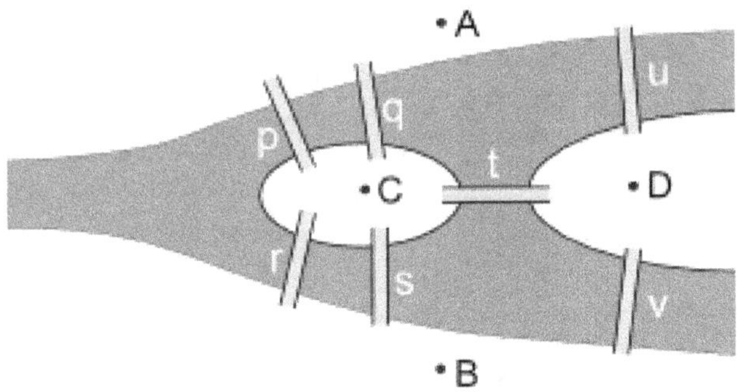


Figure 3

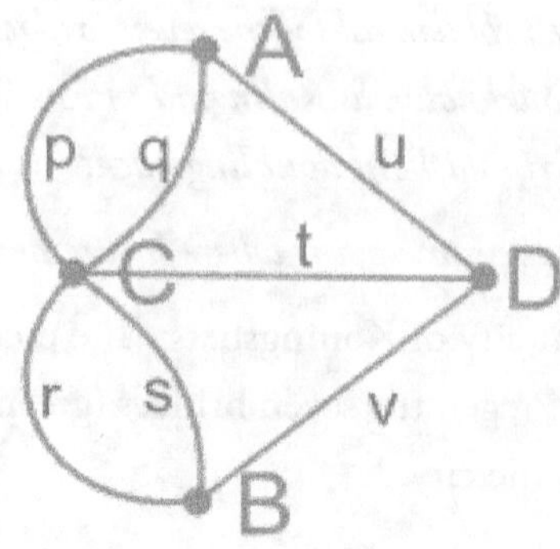

Figure 4

Coming down to brass tacks, can you draw each line *p, q, r, s, t, u,* and *v* only once, without removing your pencil from the paper?

(please note that you can start your journey from any point)

Please try.

Successful?

In case this is confusing, we can make the task easier by starting with simpler, more familiar shapes (Figure 5)

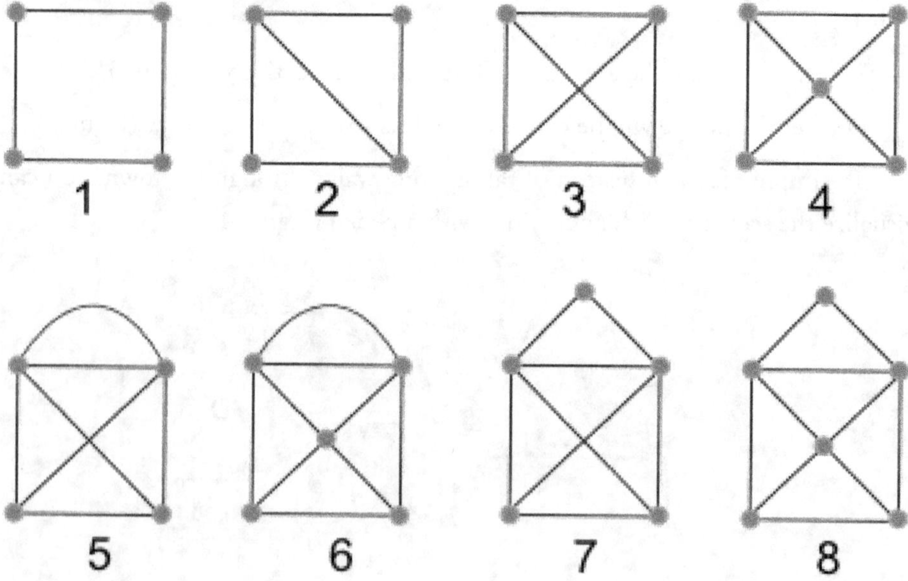

Figure 5

Remember to draw all the lines connecting the blue dots in the 8 rectangles but *never* go over any line more than once, and *do not* remove your pencil from the paper.

We need to define certain terms to be able to formulate an answer to our original question - Is it a feasible task or is it impossible?

To forge ahead we need to learn a few special words:

A point (blue dots in Figure 5) is called a *vertex* and multiple blue dots in the picture are *vertices*.

A line on any of the 8 rectangles, horizontal, vertical, diagonal or sloping is called an *edge*.

The number of edges that lead to a vertex is called a *degree*.

The whole diagram (rectangles 1 to 8) is referred to as a *graph*.

A route around a graph that visits *every vertex* once is called a *simple path*.

A route around a graph that visits *every edge* once is called an *Euler path*.

Examples:

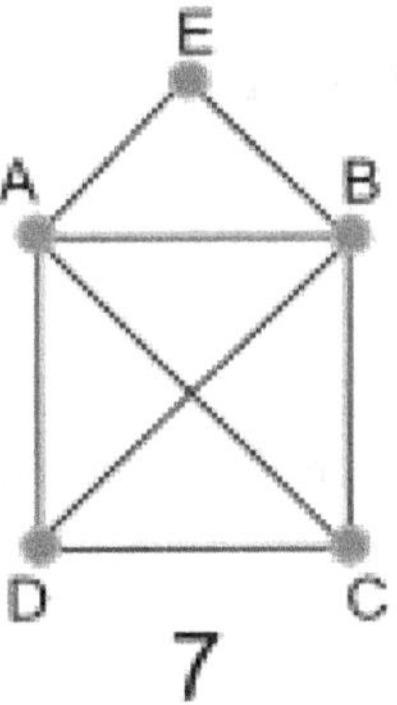


Figure 7

The figure above has

- 5 vertices A, B, C, D, and 8 edges AB, BC, CD, DA, AE, BE, AC and BD

- Vertices A and B have degree 4

- Vertices C and D have degree 3

- Vertex E has degree 2

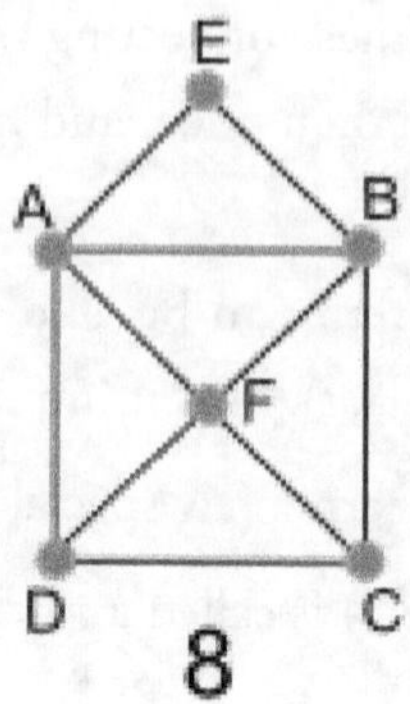


Figure 8

The diagram above has:

6 vertices A, B, C, D, E and F

10 edges AB, BC, CD, DA, AF, BF, CF, DF, AE and BE

Vertices A, B and F have degree 4

Vertices C and D have degree 3

Vertex E has degree 2

Imagine the lines are bridges. If you cross them only once the puzzle is resolved.

So we need an *Euler path*.

Through repeated observation it has been found we can identify the graphs that have an Euler path by the number of vertices *that have an odd degree.*

Vertices *that have an even degree fail to secure an Euler path!*

## SUMMARY:

The number of vertices of odd degree *must be either zero or two.* If not there is no Euler path.

If there are two vertices with odd degree then they are the *starting* and *ending* vertices.

Why?

Because a path leads into a vertex by one edge and out by a second edge. So the edges *should* come in pairs or as *even numbers*. Only the start and end point can have an *odd degree.*

What has all this got to do with the Seven Bridges of Koningsberg with which we started?

Please refer back to Figure 4:

Vertices A, B and D have degree 3 and vertex C has degree 5. This graph has 4 vertices of *odd* degree. So it does not have an Euler path.

The conclusion is that it is *impossible* to visit the seven habitable areas of Koningsberg by crossing the seven bridges *without doubling back!*

This is how Leonhard Euler solved the problem 3 centuries back.

Difficult to understand?

Do not worry - you are not alone.

You fall in the *majority group!*

## NOTES:

Leonhard Euler (1707 – 1783) a Swiss national is considered by many in the scientific community as one of the greatest mathematicians of all time. He is the author of the *Archimedes Constant,* the ratio of the circumference of a circle to its diameter commonly known as *Pi.*

**Illustration 20:** Leonhard Euler - Shutterstock _1486106429

# PART III

## KNOWLEDGE FRONTIER AREAS

# Thirty Two

# The Big Bang Revisited

*Science works on the frontier between knowledge and ignorance. We are not afraid to admit what we do not know - there is no shame in that. The only shame is to pretend that we have all the answers.*

**Neil deGrasse Tyson (1958 - ) American Astro physicist**

We are poised to examine strange aspects of our universe especially our home planet Earth. The scientific community is unrelenting in its efforts to uncover the truth through observation of natural events and their conformity with laws created by humans considered as basic and *axiomatic.*

There have been notable successes in following this path but the end point has remained elusive. It does seem that the final revelation cannot arise through human intellect alone but transcends into the realm of *God, Religion and Spirituality.* On this dilemma, humankind is divided into two camps the *Creationists* and the *Evolutionists.*

The quest for the answer to the riddle of the universe seems unending. In the narrative that follows we review the story thus far.

*If everything in the universe evolves towards increasing disorder, it must have started out in an exquisitely orderly arrangement. This chain of logic, purporting to explain why you cannot turn an omelette into an egg, apparently rests on a deep assumption about the very beginning of the universe. It was in a state of very low entropy, very high order. Why did our universe pass through a period of such low entropy?* American theoretical physicist, Sean M Carroll raises this legitimate question.

Explanation?

Part I, Ch.2 of my earlier book *Our Universe An Unending Mystery* enumerates the main features of the Big Bang Singularity. The present narrative seeks to take this arcane thought process to a higher level. An answer to Carroll query would find a place as we go along. Meanwhile there are questions intrinsic to the Big Bang phenomenon that remain unanswered:

- How could such a mind boggling colossal universe in evidence today emerge out of virtually *nothing?*

- What was the *reality* before the Big Bang?

- Is our universe a standalone phenomenon or one of many possible universes or *multiverses?*

The present analysis seeks to probe some of these issues for possible answers. In a nutshell -

*Is the present Big Bang Theory final and immutable,*

or

*With advances in theoretical physics in the 21ˢᵗ century, likely to be super ceded by an alternative theory with modified fundamentals?*

The more curious we get about the great cosmic unknowns, our solar system and planet Earth in particular, more questions pop up for which there are no answers.

Inquiring about the nature of anything - where it is, where it came from, and how it came to be inevitably leads us back to the primal question -

*The nature and origin of our universe and everything we find in it.*

No matter how far back we go, this lingering question always remains. At a certain stage it dawns on the researcher, that the entities believed to constitute the starting point of the universe did not necessarily exist - if so how did they come to be?

Simultaneously, the same teaser question - *How could something so prodigiously colossal emerge from nothing?*

We do not have an answer that can hold up to scientific scrutiny. So are we totally babes in the wood on such fundamental questions or is there something about our universe that we *do know and are sure of?*

This time the answer is more reassuring.

Our universe, based on our current understanding of it stands verified to reveal it is made up of *matter* (<5%), *dark matter* (27%), and *dark energy* (68%). Surprisingly, the breakdown reveals that we know very little about 95% of our universe, the *dark part!*

In respect of dark matter and dark energy we do know that they exist based on their incontrovertible observed effects on cosmic structures in the intergalactic universe. This inference pertains to the macro world. At the micro end of the spectrum we rely on the Large Hadron Collider type simulation studies of the Big Bang Singularity that cannot be considered as conclusive evidence. However, majority of the scientific community regard the Big Bang Theory its shortcomings notwithstanding to be the best explanation we have for our universe. It holds up better to scientific scrutiny than alternatives like the *Steady State Universe* or *Many Worlds Interpretation.*

A review of factors that support the Big Bang Theory:

- Telltale fluctuations in the Cosmic Microwave Background Radiation following the Big Bang are in evidence even today after almost 14 billion years.

- Observations by astronomers in the last 3 centuries confirm the formation and correlation between large scale cosmic structures in our universe like distant stars and galaxies as predicted by the Big Bang Theory.

- Gravitational Lensing phenomenon. (Please refer to *Our Universe An Unending Mystery,* Ch.16,pp.131 General Theory of Relativity for details)

- Finally, had our universe always existed as visualized in the Steady State Universe Theory an *infinite* number of lifetimes would have passed before reaching the present. This is not true. You are at *this very moment* reading this book. Aside from the above reasoning, Big Bang Theory predictions are scientifically verifiable in the present day e.g.our universe is made up of matter, obeys the same laws of Classical and Relativistic physics everywhere except within the atom where they are super ceded by quantum laws.

At the cost of repetition let us review the main features of the BB Theory:

Our universe emerged from an incredibly *dense and hot dot* 13.8 billion years ago. Ever since it has been expanding, cooling and *gravitating*.

It consists of dark matter, dark energy and a small segment of normal matter, fundamental particles called *neutrinos*, (explained in a later chapter entitled *Weird Particles*) and radiation.

Today, it has myriads of galaxies, stars, planets, heavy elements, and in one location (planet Earth in the Milky Way Galaxy) intelligent and technologically advanced human life. These structures did not exist in the beginning. They arose as a result of cosmic evolution.

Based on such a fundamental premise, scientists of the 20th century reconstructed a timeline of how our early universe consisting of hydrogen and helium, transformed to the complex element rich structure of today. They could have tracked the journey in reverse - starting 20th century, peeked backwards in time, and surmised where an individual structure or component came from. For each answer they got, they could have dug deeper and asked, *How did that arise where did that come?*

A little thought reveals that such lines of questioning, forward or backward are endless. *We do not have all the answers, at least not yet!*

Be that as it may, it is still exciting to track the story of how our universe evolved over gargantuan swathes of time. The reality of today is that our universe emerged from complex molecules comprised of atoms in the *periodic table* i.e. the raw ingredients that make up the building blocks of matter we have in our universe. These ingredients evolved over gigantic timespans. Multiple generations of stars lived and died, with products of their nuclear reactions recycled into future generations of stars. Without such happenings planets and complex chemistry of today would not have been possible. Supportive evidence manifests in *supernova remnants* (powerful, scintillating explosion of stars) and planetary *nebulae* (gigantic clouds of dust and gas in between stars). Such celestial events enabled stars to recycle their burnt heavy elements back into the interstellar medium and give birth to the next generation of stars and planets. Extending this chain of logic we can infer that these processes evolved in two ways. First, the heavy elements necessary for chemical based life were created and second, *more importantly* enabled the emergence of the *human life form* endowed with intelligence to probe and enlighten how it all happened in the first place. Summarizing yet again,

In order to form modern stars and galaxies, we surmise that the following sequence of events occurred:

- Gravitation pulling small galaxies and star clusters into one another.

- With increasing gravitation, simultaneous creation for the first time of an equal mix of normal matter and *antimatter*, dark matter, dark energy, radiation (CMBR) and abundant light elements. A corollary to the above reasoning is that every reaction observed could only create matter and *antimatter* in equal amounts.

  But the universe we have, despite beginning in an incredibly hot and dense state where both matter and antimatter could be created in copious amounts, must have had some way to generate a matter/antimatter *asymmetry* where none existed initially. Some of the more obvious ways in which this could have occurred are:

- An out of equilibrium set of conditions, that naturally arose in an expanding, cooling universe,

- Generation of sufficient *C violation* (charge conjugation symmetry violation) and *CP violation* (charge conjugation and parity symmetry violation) resulting in our universe generating *something from nothing*.

(*Our Universe An Unending Mystery* -

CPT Invariance, Ch.19.p.144)

To this day we are *not certain how* all this actually happened. An associated mystery relates to the viable ways that could have generated dark matter.

We know after extensive theorizing that whatever dark matter is, it cannot be composed of particles that are present in the Standard Model. Whatever its true nature it requires physics beyond our knowledge frontier.

Be that as it may, we can logically deduce that in all cases it would require an enormous amount of *energy*. This begs the question - Where did this energy emanate from?

We cannot be sure but possible sources are:

*Pre* Big Bang: an infinitely hot, dense, unstable, orderly state with zero entropy leading to *cosmic inflation followed by reheating* a popular concept in scientific circles. What *actually* happened subsequently provides us a choice between two options:

- The *Lady Gaga* route, claiming it must have been *born this way*. This is the easier way out since it precludes further discussion and debate. A theoretical physicist would call this approach *giving up*.

- The alternative is what physicists do best create an *Imaginary Route:* Concoct a theoretical mechanism that could explain the initial conditions, followed by fleshing out concrete predictions that differ from the conventional Standard Model forecast.

Cosmic inflation came about as a result of taking the second approach, and it literally changed our conception of *how our universe came to be.*

Let us explore this option further:

*Exponential Expansion* took place during inflation - an all pervasive, mind bogglingly relentless phenomenon.

With every ~$10^{-35}$ seconds that passed, the volume of any particular region of space doubled in each direction, causing particles and radiation to dilute and inherent curvature to quickly become indistinguishable from flat.

Instead of extrapolating *hot and dense* back to an infinitely hot, infinitely dense singularity, inflation followed by reheating suggests that *perhaps the super hot Big Bang was preceded by a period where an extremely large energy density was present in the fabric of space itself, causing the universe to expand at a phenomenal rate, and when this inflation phase ended, a reheating phase followed, that energy got transferred into - matter, antimatter, and radiation, creating what we see as the hot Big Bang. By this reasoning, the aftermath of inflation and reheating.*

So what have we got?

Our universe was created with the same temperature everywhere, spatial flatness, and with a spectrum of density fluctuations, that seemed to merit further observation and study.

From just empty space comprising *quantum field energy* the process created the entire observable universe, rich in structure, as we see it today.

This inflation/reheating based reasoning behind getting our universe from nothing, is not considered to be good enough to satisfy all researchers. One shortcoming is that even in space of total void, quantum fluctuations inherent in the quantum field theory governing fundamental interactions cannot be removed. Accordingly, as inflation progressed from the earliest stages, these fluctuations got stretched across the universe giving rise to seed density and temperature fluctuations. The credibility factor for such occurrences is high since they can be observed even today. For most researchers, space and time integral to our present day universe, the laws of physics, the fundamental constants, and some non zero quantum field energy intrinsic to the fabric of space itself, is very much divorced from the idea of nothingness.

Are we back to square one?

Not quite.

We can still imagine, a location outside of space; a moment beyond the confines of time; a set of conditions that have no physical reality to constrain them.

And these imaginings - if we define the physical realities as things we need to eliminate to obtain true nothingness are certainly valid, but only philosophically!

That is the difference between philosophical nothingness and a more physical definition of nothingness. There are in fact four scientific definitions of *nothingness*. Probing this topic further would mean entering an area that is complex and counter intuitive - in short a major digression that we wish to avoid.

We can however based on the analysis so far state that without a physical theory to describe what happens outside of the universe, beyond the realm of physical laws, the concept of true nothingness is physically ill defined.

Fluctuations in spacetime itself at the quantum level get stretched across the universe during inflation, giving rise to imperfections in both *density* and *gravitational* waves.

While inflating space can justifiably be called *nothing*, it is a debatable point.

Therefore, within the thinking realm of the theoretical physicist, it is impossible to accept the idea of *absolute nothingness.* What does it mean to be outside of space and time, and how can spacetime predictably emerge from a state of non existence? If we pursue this thought process, it leads to the related question - Where do the rules of the *micro world* that govern quanta - the fields and particles arise from?

We are then required to assume that space, time, and the laws of physics themselves are not all pervasive and eternal, when in fact they *may ultimately turn out to be so.*

The opposite view is that if we accept the physical definition of *nothing* as valid, then the universe seems to have arisen from nothing and misgivings about our cosmic origin vanish.

Unfortunately, inflation by its very nature cannot be *verified* since it erases telltale information that may have got imprinted from a pre- existing state in our observable universe. We are forced to conclude that the arguments so far have led us into a state of suspended animation from which there is no escape route! In view of these arguments it seems fair to state that the validity of *nothingness is and will always be a mere construct of our minds and remain an unknowable entity.*

There is little doubt that the quantum world is a *spoiler* but for what it is worth would we have been capable of a full explanation in an exclusive classical world?

For starters, mind bogglingly colossal matter arising out of nothing, violates the First Law of Thermodynamics (matter can neither be created nor destroyed). Big Bang Theory apologists would counter this stating the theory deals with the *evolution* of the universe, not its *creation.* They would resort to *circular reasoning* to explain the impasse stating that as you approach the point of creation of the universes the Laws of Classical Physics (Laws of Thermodynamics) tend to break down.

Specifically, the Second Law of Thermodynamics (Law of Increasing Entropy) adopts the following line of reasoning - Our universe is constantly losing usable energy and never regaining it. This cannot go on forever. Hence our universe *cannot be* eternal. It began at a zero entropy point, and since then has been winding down with the entropy level increasing. This leads to the question - *Who wound the clock to zero entropy in the first place?*

This question would excite the theologian since it points to a *Creator*. Astronomers on the other hand would prefer *Evolution* as a more plausible explanation. In this context Albert Einstein had stated - *For every one billion particles of antimatter there were one billion and one particles of matter. And when the mutual annihilation was complete, one billionth remained and that is our present universe.*

Since the words emanate from one of the greatest physicists the world has ever known we need to accept it as the *last word* in our present context.

The summary concludes that from the angles of Classical and Quantum physics introduction of the concepts of Cosmic Inflation and Reheating prior to the Big Bang Singularity are well reasoned and legitimate. Unfortunately, it has left seemingly unanswerable questions in the mind of the space researcher.

The earlier Einstein quote on matter and antimatter provided some relief from the dilemma but related questions on *energy* remained shrouded in mystery.

For further insight, we have to depend on theory since the conditions of the Big Bang have not been replicated in a laboratory in terms of measurement of infinitesimal *sizes* and *intervals of time.*

Einstein's quote implies that matter and antimatter cannot coexist. They annihilate each other *instantaneously* releasing high energy radiation. We can

accept this explanation since it does not conflict with the Laws of Conservation of Mass & Energy, the building blocks of Classical Physics. Furthermore it is borne out by proven occurrences like:

- Radioactive decay and the atom bomb as examples of *nuclear fission at different speeds* depicting the conversion of mass to energy.

- The chemical reactions in respect of events following the Big Bang are examples of the reverse i.e. the conversion of *energy to mass*.

(Matter and antimatter particles would also have opposite *quantum numbers* that relate to *quantum spin*. We The remind the reader that the quantum world is the *Wonderland of Alice* with its own set of incredible laws).

- Regarding the *Inflation Phase* following the Big Bang, matter first formed and decoupled from energy at $10^{-35}$ of a second after the Big Bang when the primordial temperature cooled sufficiently for energy to convert to matter. This is an example of nuclear fusion. Cosmologists refer to this seminal stuff as baryonic matter that is observable. They opine that in addition to the baryonic matter, a much larger quantity of invisible, unobservable dark matter was also formed. Certain aspects relating to the *measurement* of infinitesimally tiny sizes and time intervals in the quantum world are highlighted to enable a realistic perception of conditions immediately *before* and *after* the Big Bang. This will enable the reader to grasp the yawning gaps that exist *even today* in our understanding of the Big Bang Theory.

For the lay person measurement of time intervals less than a second is difficult to comprehend. In common parlance we often say - *it happened so quickly, in a split second* or *in a fraction of a second.*

We cannot imagine entities tinier than one fourth or one eighth of a second. In the Big Bang Theory we are talking of time intervals measuring 0.00000......1 i.e. 35 zeroes preceding 1 second indicating unimaginable *smallness*. After the passage of this incredibly tiny interval of time following the Big Bang, the universe was no longer energy alone, but a combination of energy and matter. It seems that the *divine* purpose was that only energy should exist even after temperatures had cooled sufficiently to permit matter to decouple from the energy flow. By an unexplainable quirk when matter and antimatter particles were evolving neck and neck after the first billion was reached, the matter particles surged ahead of the antimatter particles by *just one*. We know this is true through a remarkable piece of *reverse engineering.*

The described excess pattern continued till the entire matter in our present day universe came into being. It continues to form in this manner to the present day and beyond - relentless and unstoppable. Little wonder that it constitutes *less than 5%* of the whole.

(For further details please refer to Part I Chapter 2 of my book *Our Universe An Unending Mystery* published by *Create Space* in 2017)

The earliest occurrences in our universe in respect of *time evolution, size expansion and temperature drop* constitutes the hallmark of the embryonic quantum world. They are counter intuitive and mind bogglingly different from numbers in our present day to day world. The Big Bang Theory is *not* an outright winner but a mixed bag. As stated earlier, a majority in the scientific community are of the view that it is the best available explanation for the history of our universe. The reality is that despite the phenomenal progress in the 21st century in the fields of cosmology and astrophysics we still do not have answers that can hold up to rigid scientific scrutiny.

*Could it therefore be the handiwork of a Creator?*

I have dealt with this aspect extensively in a later essay on *God and Religion* that will provide the reader a full perspective.

In respect of the Big Bang Singularity, it is useful and timely to regroup our thoughts on this seminal event believed to have occurred 13.8 billion years ago.

Singularities are real but *forbidden zones* that defy our current understanding of physics. They fall in the *Knowledge Frontier Area* because their definition is incomplete without use of the word *infinity*. Infinity is the delight of the mathematician but bugbear of the theoretical physicist. Where does a singularity come from?

*We do not know.*

Regarding momentous subsequent events involving mind boggling changes in *time* (micro), *size* (macro) and *temperature* (macro) stretching from the beginning to the present day and beyond, the note worthy are:

- Emergence of *life forms* with the *human life form* at the pinnacle endowed with intelligence to *observe, record,* and *interpret* occurrences in our universe. Also to *refine and improve* them from time to time as an ongoing process that could turn out to be unending.

- Realization that we are creatures living on a unique planet, circling a star clustered together with several hundred billion other stars in a galaxy soaring through the cosmos.

- This panorama is unfolding within an expanding universe that began as an infinitesimally tiny singularity that appeared out of nowhere for unknown reasons. Our attempts to visualize these events have led us to imagine an infinitesimally tiny balloon expanding to the size of our present day universe that continues to expand and accelerate perhaps forever. The imagination can be stretched further. The balloon analogy is our best bet because *space* and *time* we take for granted today *did not exist prior to the Big Bang.* In other words, the Big Bang Singularity did not occur *in* space; rather, space began inside of the singularity. Prior to the singularity *nothing* existed - no space, no time, no matter and no energy. Is this reality applicable for the *past, present,* and *future? We do not know.*

On an overall review, we can finally conclude that The Big Bang Singularity Theory is the best explanation we have *today* for the origin of our universe and everything that we are able to observe in it.

*From unimaginable smallness expanding to an equally unimaginable large size through cosmic processes we describe as 'Singularity' and 'Big Bang' occurring in between.*

*Celestial bodies so created move in defined orbits at prodigious speeds. In attempting to unravel these events, the human mind is overwhelmed by shock and awe.*

*Had human life not emerged through Evolution, these wonders would have forever remained hidden.*

*Location of this occurrence on an inconspicuous planet Earth in a very ordinary galaxy, the Milky Way defies a definitive analysis whether the underlying cause is Accident or Design.*

*Our Universe of the 21ˢᵗ century, despite significant advances in science and technology has in many respects remained indecipherable.*

*Efforts by man to breakthrough such knowledge frontier areas continue relentlessly. Hopefully success will come sometime soon and not remain an elusive dream for all time.*

*In the interim man needs to shift focus from his present pursuit of acquiring knowledge of the unknown through the intellect alone and seek solace in the*

*God entity to reveal the unknown.*

*The two paths should be seen as complementing rather than negating each other.*

We conclude this essay with random thoughts on the revolutionary theory of *Plasma Cosmology.*

Author Eric Lerner proposes that the *Big Bang Never Happened!*

The reader will recall that of the theories doing the rounds as possible alternatives to the BB Theory, the BB had an edge mainly because of the existence even today of the CMBR.

Lerner points out as late as 1980 of the possibility of the alternative Plasma Cosmology Theory as proposed by *Hannes Alfven.* There is the school of thought that that the PC Theory is zilch, and yet is gradually gaining ground in scientific circles as an alternative to the BB Theory. It suggests that the Universe *did not* have a beginning and that the cosmological red shift is *not* due to movement based Doppler shifting of light. If therefore the universe is not actually expanding as proposed by the Big Bang Theory *but only appears to be* then we are back to square one.

The story of the universe continues to confound us.

*It seems it is one of those factors we will never get to know!*

# Thirty Three

# Weird Particles

*Everything we call real is made of things that cannot be regarded as real.*
*If quantum mechanics has not profoundly shocked you, you have not*
*understood it yet.*

**Niels Bohr Danish Theoretical Physicist (1885 – 1962)**

Niels Bohr was referring to the wonders and surprises of the *micro or quantum world* in the Quantum Era. This alert to the reader is both timely and apt as we get drawn into new depths of the brave new *Wonderland of Alice.*

For the first time, theoretical physics is governed by counter intuitive laws called *quantum mechanics.*

For a full perspective we need to briefly review main events of the 2 centuries long Atomic Era that preceded the Quantum Era, from the mid eighteenth century featuring John Dalton (1766 - 1844) to Niels Bohr (1885 - 1962)

Bohr had a presence in both the eras. He is best known as a pioneer in quantum mechanics alongside stalwarts Erwin Schrodinger and Werner Heisenberg.

Be that as it may, in the quantum era the scientific community shifted its focus to the tiny end of the size spectrum i.e. *smallness* as opposed to *vastness* of the universe.

In this context two mind blowing discoveries in 1924 and 1968 revolutionized our thinking on the size of our universe in respect of *extreme vastness* and *extreme smallness.*

Till 1924, we believed that our parent *Milky Way Galaxy* constituted our entire universe. By earthly standards, the near circular spiral MWG with a diameter measuring over 100,000 light years seemed prodigiously huge. American astronomer Edwin Hubble with the aid of the *Hubble Telescope* established in the period 1920 - 1924 that the MWG is an inconspicuous speck, just one of the estimated 125 *billion* galaxies in our universe. Edwin Hubble's arrival heralded an entirely different assessment of the *size* of our universe. Its outer periphery seemed to expand with acceleration in every direction.

It triggered the thought process that sometime in the future, the only stars visible to earthlings would be *those in the Milky Way Galaxy merged with neighbouring galaxy Andromeda, presently 2.5 million light years away. Gravity would cause them to hurtle towards each other and merge in an awesome Cosmic Dance. Following merger, light from the remaining 125 billion galaxies would not be able to move fast enough to compensate for the accelerating expansion of our universe and will never reach us.* We face the prospect of living in a single merged Galaxy surrounded by an *ocean of unfathomable darkness!*

When?

*Five billion years from now.*

At the other end of the size spectrum the widely held Classical view that the atom is the limit of smallness with unique properties for each element in the *Periodic Table* came under scrutiny with the emergence of subatomic particle physics in 1968.

In this essay we are focusing on the lower end of the size spectrum i.e. the limits of *smallness* and its implications.

What prompted the scientific community to review whether the atom is in deed the smallest entity of an element in the Periodic Table?

That it consisted of a heavy central nucleus comprised of protons and neutrons, and a relatively *distant* periphery of massless orbiting electrons?

The answer lay in understanding the structure of the atom revealed by experiments conducted earlier in the Atomic Era by Dalton, Thomson and Rutherford.

Against this background, in 1968 *electron proton scattering* experiments conducted at the Massachusetts Institute of Technology in collaboration with the Stanford Linear Accelerator Center revealed for the first time that protons and neutrons a.k.a *nucleons* within the atomic nucleus have an inner structure. The study revealed that they are made up of yet smaller microscopic particles called *quarks* that cannot be broken down any further. Quarks replaced atoms as the smallest unit of mass. This development marked the commencement of a whole new body of knowledge called *Particle Physics*.

I digress from the realm of physics with an analogy from English literature that fits our present context......

Particle Physics represents the threshold of a brave new world reminiscent of the *Wonderland of Alice* created by Charles Lutwidge Dodgson the English writer better known as *Lewis Carroll*. He was famous for his extraordinary ability in the fields of word play, logic and fantasy.

*Yes* you read that right. The particle world is a world of *fantasy*!

I quote portions from Lewis Carroll's masterpiece *Wonderland of Alice* that throw light on the fantasy aspect of the quantum world.

*Alice is sitting with her sister on the riverbank and is very bored. Suddenly she sees a White Rabbit running by her. He is wearing a waistcoat and takes a watch out of it, while muttering to himself 'Oh dear! Oh dear! I shall be late!' Alice gets very curious and follows him down his rabbit hole.*

In our narrative that is what we are about to do - *Enter the rabbit-hole of the quantum world.*

Alice?

*She decides that she must have been changed into another girl in the night as she cannot remember her multiplication tables or geography correctly and is not able to recite a poem properly.*

As in Wonderland of Alice, we have to enter the subatomic quantum world with a simultaneous lapse in memory of Classical Physics and suspension of disbelief in respect of the laws that govern Quantum Mechanics.

Getting back on track, in the quantum world scientists have divided quarks into six types named whimsically as -

*Up, down, strange, charmed, top* and *bottom.*

Aside from quarks being of a different genre, there was some activity taking place within the nucleus of an atom that was keeping it stable e.g. the nucleus of the element helium has 2 positively charged protons and 2 neutral neutrons. As per Maxwell's Classical Electro Magnetic Theory 2 positively charged protons within the confined space of a helium nucleus would strongly repel each. The helium atom would be unstable. We know from experience that the helium atom *is stable.* Researchers concluded that the electromagnetic repulsion within the helium nucleus is being neutralized and overcome by a stronger reverse force.

They called this the *Strong Nuclear Force* the strongest of the 4 *Fundamental Forces* that account for the multifaceted movements occurring in our observable universe from the *micro* level of the nuclei of atoms to its extreme cosmic *macro* edges.

The strong nuclear force is in fact 137 times more powerful than the next in line fundamental force, the electro magnetic force that we are familiar with as kids picking up metallic objects with toy magnets. The ratio 137: 1 is called the *Fine Structure Constant,* one of the most important constants in modern physics.

(Had it been even slightly different say 139: 1 or 135: 1 it would have caused manifold disturbances in the stability of our universe from the micro level of the atomic nucleus to its extreme macro periphery)

Regarding the strong nuclear force, we use an analogy from our day to day lives to illustrate how this force is generated.

The layperson's understanding of a force is of a *push* or *pull* effect on an object. An example is the Indian housewife who moves a wooden roller up and down on a ball of dough to convert it to a *chapati.* What is happening to the protons and neutrons within the dough as it flattens out?

Explanation for this innocuous activity leads us to probe the inner constituents of the atom i.e. the even smaller matter particles, the quark and the *lepton* together with the force mediating particle the *gluon.* The reader is likely to get confused with this business of particles and their unfamiliar names like quark, *lepton* and *gluon.* This is the start of our particle journey, the proverbial tip of the iceberg. There is more to come.

The *strong nuclear force* arises as a result of the quarks within the protons and neutrons playing an endless game of *catch* with the massless gluons which are called exchange or force mediating particles. An analogy from cricket might help in understanding what is going on. Imagine a cricket stadium where players are in the midst of a fielding practice session. The wicket keeper and first slip are in position practicing taking sharp catches. They throw the ball at each other *continuously* without a break. Put differently, they play a continuous *game of catch* with the cricket ball. Any tendency of the two players to move away from each other (electromagnetic repulsion) is kept at bay by the much stronger urge (strong nuclear force) *not* to interrupt the catching practice between them. This is a *macro* version of the strong nuclear force prevailing over the electromagnetic repellent force within the confines of the nucleus of the helium atom. The same action explains the stability of the nucleus of *most* of the elements on planet Earth.

Reverting to the helium nucleus (2 positively charged protons ($p_1$ and $p_2$ + 2 neutral neutrons, $n_1$ and $n_2$) we can meaningfully theorize as follows:

Each proton is made up of 3 quarks ($q_1$, $q_2$ and $q_3$ *with mass* and 3 exchange particle gluons ($g_1$, $g_2$ and $g_3$) *without mass.* Each gluon is being endlessly exchanged with a quark pair i.e. there are 3 endless exchanges taking place within each proton (3 *games of catch* by the quark pairs within each proton. The possible combinations are,

$q_1$ and $q_2$ play a game of catch with $g_1$, $q_2$ and $q_3$ with $g_2$, and $q_3$ and $q_1$ with $g_3$.

*The point to remember is that the strong nuclear force, the strongest of the four fundamental forces depends on these catching games being ever lasting!*

In addition to the reactions within each proton, many other combinations are possible e.g. adjacent protons can react with each other and with adjacent neutrons generating more *ever lasting* games of catch.

These endless catch games between protons and neutrons requiring their respective quarks to remain glued to their positions is the best explanation we have for the existence of the *strong nuclear force.*

For a full understanding, we need to probe further on the nature of quarks, leptons and gluons.

QUARKS: *Quantum Colour.*

The pioneers of particle physics realized in their early studies that all quarks are *not* of the same type.

As mentioned earlier, they can be divided into six types or *flavours - Up, down, strange, charmed, top,* and *bottom.* Besides, they have different *colours.* Colour in quantum mechanics refers to *charge* and not *hue* as in common parlance. Rules govern the permissible combinations of quarks and gluons based on their colour charge akin to a secret language *only they can understand.* The end result is that quarks and gluons are the *only* subatomic particles that play the endless game of catch and thereby experience the strong nuclear force.

*Quantum spin.*

The spin concept keeps coming up frequently when we probe subatomic particles. We need to understand its significance in respect of the matter particles, quarks and leptons on the one hand and the force mediating particles, gluons on the other.

Matter particles are called *Fermions* (named after American theoretical physicist Enrico Fermi) and Force Mediating particles *Bosons* (after Indian mathematical physicist Sathyendranath Bose). Fermions have *half integral* spins or multiples thereof whereas Bosons have *full integral* spins.

What does this mean?

As a simple definition we can compare a *spinning* subatomic particle to a kids spinning top. The reality is far more complex but we are refraining from providing details as it is beyond the scope of this essay. Suffice it to say that quantum spin is an aspect of the general strong *rotational symmetry in nature.*

We next highlight the significance of subatomic particle groups that fit into the broad subject of *Particle Physics.*

## HADRONS, MESONS AND BARYONS

*Hadron* has a broad connotation that covers *all* quarks.

*Baryon* is a type of hadron that contains three quarks and in quantum terminology considered *heavy.*

*Meson* consists of one quark and one *antiquark* considered to be of *intermediate mass.*

Regarding the hadron, meson and baryon groups we make it easier for the layperson's understanding through an analogy from our day to day lives. The hadron can be compared to the entire *avian* species (atoms), the baryon to *large predator eagles* (protons and neutrons) and the meson to *small sparrows.*

## ANTIPARTICLES

An antiparticle is the antithesis of its corresponding particle of the same mass but is of an opposite charge e.g. an electron's antiparticle is a *positron.* When they collide, they annihilate each other and release a stream of their exchange particle the *photon.*

## UNPARTICLES

A group of particles unexplainable by the Standard Model of Particle Physics.

In the SM, matter particles exist in states defined by energy, momentum and mass e.g. electrons always have the same mass regardless of their energy or momentum. In contrast, the exchange particle, photon can exist with their mass, energy and momentum properties scaled up equally i.e. they are *Scale Invariant.* Such particles are in a sense notional and accordingly called *un particles.* Example of an unparticle is the *neutrino.*

Do notional unparticles provide a clue to the existence of a *fifth fundamental force?* A plausible answer to this query is believed to lie in the behaviour of the sub-atomic particles called *muons.* These minuscule entities are similar to the electron but 200 times heavier. Scientists found muons to *wobble* much more than they should when sent around a a 14 metre ring to which a magnetic field is applied. This research topic hints that there is a force of nature unknown to science that influences muons. This fifth fundamental force could possibly explain some unresolved questions like the acceleration in the expansion of our universe. Presently this phenomenon is sought to be explained as the result of the mysterious *dark energy*. Instead it could be the existence of a fifth fundamental force.

Because of its heavy mass, the muon was initially thought to be the particle predicted by Japanese physicist Yukawa Hideki in 1935 to explain the strong force that binds protons and neutrons together in atomic nuclei. It was subsequently discovered that it is more appropriate to assign it as a member of the lepton group i.e. it reacts with the electromagnetic, weak nuclear force and gravity *not* with the strong nuclear force. The muon is relatively unstable with a lifetime of only 2.2 *microseconds* following which it decays by the weak force into an electron and two kinds of neutrinos. This establishes the link between the muon and neutrino and possibly with a fifth fundamental force yet to be discovered.

## QUASI PARTICLES 1 & 2

There are a large number of exotic fundamental particles, consequences of *super symmetry* in this category. Each of them has intriguing properties that open up new vistas of investigation, a fertile field for the budding particle physicist. A few other promising particles with hints what they might lead us to are given below.

*Axion* - An extension of the Standard Model postulates it as a tiny, partial successor to the photon. It is a notional particle believed to be almost massless (a millionth of an electro volt) about a *trillion times lighter than the electron*. Only weakly interacting with other particles, the axion is a possible building block of *Quantum Chromodynamics*, Black Holes, and Cold Dark Matter in our universe.

*Wimp* - Weakly Interacting Massive Particle emerged into the limelight in 1970/80 and is a possible cousin of the axion. Again notional, they do not absorb or emit light or interact with other particles. On colliding with each other they self destruct and transform into high frequency *gamma ray radiation*. The wimp is a consequence of super symmetry and fits into the role of super partner to a known particle. It is a top contender for mysterious dark matter. Believed to have been formed at the dawn of creation, the Big Bang, many may still be around. When a credible estimate of their number becomes available that matches the amount of dark matter revealed through experiments, it would herald the arrival of the *Wimp miracle*. Scientists are hoping for a conclusive result in the next 10 -15 years.

*Fracton, Gluino, Wino, Photino, Neutralino, Chameleon and Higgsino*

*Fracton*

The theoretical possibility of the Fracton was first realized by physicists as recently as 2011. They are exotic quasiparticles that are either totally immobile or move only in a limited way when compared to their normal behaviour pattern. Scientists have attempted to convert this unusual property of the particle into a commercial opportunity in the working of quantum computers.

At Caltech, Jeongwan Haah and his team developed the *Haah code* a computer algorithm based on this property. The team reasoned that the observed particle movement is *illusory*.

Consider how a scientist observes an electron to move. The electron *moves* because its space is filled with electron - positron pairs momentarily popping in and out of existence. In one such pair, the positron is *on top of* the original electron and they annihilate. This leaves behind the electron from the pair displaced from the original electron. Since there is no way of distinguishing between the two electrons, we perceive the illusion of a single *moving* electron. It explains the choice of *fracton* for the particle.

We need to stretch our imagination further as to *how* a stream of particles come into existence. Pairs of particles arise *not* out of vacuum but in *squares*. In such a situation, a square might arise so that one antiparticle lies on top of the original particle (like the positron atop an electron) annihilating that corner. A second square then pops out so that one of its sides annihilates with a side from the first square. This leaves behind the second square's opposite side, also consisting of a particle and an antiparticle.

The resultant movement is the illusion of a particle - antiparticle pair moving *sideways in a straight line* called a

*fracton phase* where a single particle's movement is restricted, but a pair can move easily. The Haah code is an example of the above phenomenon i.e. that particles can move *only* when new particles appear in never ending patterns that keep repeating themselves. Such patterns are called *fractals.*

Our imagination is again pressed into service. This time it is a square with four particles at the vertices. On zooming in to a particular vertex, you find a smaller square, whose individual corners will show smaller and smaller squares i.e. an endless set of diminishing squares.

Such a structure would require infinite energy for maintenance which is theoretically impossible. The result is a stable immovable unit called a qubit the foundation cornerstone of *quantum computing.* This is in variance with the classical idea of free movement of particles in a vacuum as dictated by temperature and pressure constraints.

Fractons as described above defy the classical concept in that they cannot shake off their microscopic character and conform to a continuum description.

Unlike Classical Physics, this aspect of quantum mechanics poses challenges when attempting to predict behaviour of fractons. The silver lining that encourages particle physicists to continue this line of research is the discovery of certain naturally occurring crystals with a structure *mathematically similar to fractons.* This thought process could shake the foundation of the principles on which the Standard Model of Particle Physics rests.

Scientists continue to search for the missing link that could lead to a *viable quantum gravity theory.* In this context, discovery of the *Higgs Boson* is an important landmark.

On an overall review, some tweaking of the Standard Model would be necessary to accommodate a Grand Unified Theory, the nature and extent of which could only emerge out of the discovery of yet new particles that have remained elusive so far.

A summary of *new particle possibilities* beyond Higgs are given below.

*Gluinos, Winos, Photinos and Higgsinos:*

According to the theory of super symmetry there could be a dozen particles *yet to be discovered* e.g.

each fermion could be paired with a boson and vice versa.

So gluons (a type of boson) would have *gluinos* (a type of fermion), W particles would have *winos*, photons would have *photinos* and the Higgs itself would have a counterpart called the *Higgsino*.

Theory apart, the LHC in its perpetual quest for new particles has found no evidence of such particles and have presently taken the view that they are unlikely to exist. However there have been a few encouraging signals.

In 2012, ultra rare particles called *B sub S mesons* were discovered, normally not found on earth but exist fleetingly when two protons collide at very high speed approaching the speed of light.

This is a small step forward since it only reveals that super symmetric particles would need to be heavier to come within the scope of a probe. Besides, physicists are not clear on the size and energy ranges within which such particles can be found and are still groping in the dark.

## Neutralino:

Hypothetical particles that have never been observed or detected. Believed to exist as a building blocks of a *Minimal Supersymmetric Standard Model.*

They are classifiable as Bosons and exist in four *eigen* states - with weak masses (100 GeV ~ 1TeV).

As a phenomenon they are comparable to neutrinos *not* directly observable.

## Chameleon:

The chameleon particle is a candidate for explaining dark energy that seems to drive galaxies apart in the observable universe. The theory states that the chameleon like its reptilian namesake can change form and escape detection. The difference is that *they change their mass with surroundings not their colour.*

Such exotic particles are possible indications of super symmetry in fundamental particles, dark energy, or dark matter. The list is likely to increase with advances in Particle Collider Technology leading to the discovery of yet newer particles.

## *The Higgs Boson:*

As compared to the discovery of humble particles of the Atomic Era like the electron, the Higgs boson required the creation of experimental energies not attempted earlier on earth. The *Large Hadron Collider* is considered as one of the most famous and successful scientific equipment of all time. The target was initially called the *God Particle* because of the shroud of uncertainty that surrounded it. Thanks to the untiring efforts of scientists in the 20th and 21st centuries we no longer have to take its existence on faith.

The Higgs boson is important as the last hold out particle to be checked in support of the Standard Model of Physics. Besides, it provided *mass* to *all* other particles. Everything that exists depends on the the Higgs boson.

It led to the *Higgs field* occupying every nook and corner of our universe.

Scientists theorized with trepidation considering that their earlier similar attempt when probing the speed of light of an *all pervasive universal ether* in Classical Physics was proved incorrect.

The Higgs field hypothesis drew many parallels from light but there were significant differences. In white light, all wave lengths exist within it, but the Higgs field is believed to break the mass symmetry of some particles that are otherwise symmetrically massless. If the Higgs field did exist its action would require an observable *carrier particle the Higgs boson* similar to the photon in the electromagnetic spectrum in Classical Physics.

The Higgs boson was seen as a local *excitation* of the Higgs field. An official announcement from CERN followed on March 14, 2013 *confirming the existence of the Higgs boson.* It was accompanied by a caveat that more study is required for a full understanding of the new *God particle.* Ongoing research on the Higgs boson could throw light on abstract topics like Anti matter and Dark matter. Some physicists hold the view that the Higgs boson could be *the missing link in the Standard Model in respect of Gravity and a Theory of Everything* (TOE).

# Thirty Four

# Speed Of Light

*Technologies that may be realized in centuries or millennia include: warp drive, traveling faster than the speed of light, parallel universes. Are there other parallel dimensions and parallel realities? Time travel and going to the stars?*

**Michio Kaku American theoretical physicist.**

Michio Kaku has in the above statement identified a few *knowledge frontier areas* that scientists constantly explore. It provides a clue that this quest by its very nature is *endless*. More specifically we are intrigued by the everyday phenomenon of light that occupies a special position after the life and death of Albert Einstein, the great scientific mind of 19th/20th century.

We often wonder what is so special about light, the constancy of its speed, and the formula $E = mc^2$ that everybody talks about but few understand.

Regarding the constancy of the speed of light, we know that the speed of light is 300,000 kms per sec or 186,000 miles per sec in air or in a *vacuum*. It drops by about 30% in most liquid media like water, and by a further 15% in solid media like a glass prism.

Be that as it may, whether air, water, or glass this range of speed when compared to speeds in our day to day lives is in a different league, beyond the realm of comparison.

For example, the fastest a human on earth can sprint is only about 13 metres per second. The fastest animal, the cheetah can sprint at double this speed.

In either case comparison with the speed of light whether the medium is air, liquid or solid seems meaningless.

By earthly standards we reckon the speed of light to be instantaneous.

If a central dome light is turned on at night we assume that it reaches every nook and corner of the room *instantaneously*. Another example from our day to day lives is of a lightning flash and the thunder clap that follows a few seconds later. It is an indication of the significant difference in the speeds of light and its humble cousin sound.

Why then is light singled out for discussion on speed and constancy, a prerequisite for the *Special Theory of Relativity?*

The simple answer is to facilitate the layperson's understanding of the concept. The upper speed limit applies not to light alone, but the entire *electromagnetic spectrum* from *radio* waves at the lower end of the frequency range to *gamma* radiation at the upper end. (With the visible *light* range falling in between these two extremes)

Einstein derived a law governed by one of the simplest but most powerful equations ever written, $E = mc^2$ where $E$ or energy on one side of the equation represents the total energy of a system, $m$ the mass and $c$ the speed of light squared on the other side.

The equation as a whole *equalizes* mass and energy. For the first time, Einstein created an awakening in the scientific world that mass and energy are manifestations of the same entity.

There are three inevitable consequences to this line of thought. The first is that

*Even masses at rest have an inherent energy within them.*

In the Classical Physics of Newton all forms of energy, mechanical, chemical, electrical or *kinetic* are associated with moving or reacting objects that can be made to do work like run an engine or light a bulb. Einstein's equation that *innocuous mass at rest* has energy within it was a revolutionary thought.

The second consequence is that the equation *enables the quantum of energy within mass to be precisely worked out.*

For example, for every kilogram of mass at rest fully converted to energy, the yield is $9 \times 10^{16}$ joules of energy equivalent to 21 megatons of TNT.

It was difficult for Classical physicists of the 19th century Newton era to accept this statement coming as it did from an unknown clerk in the Swiss Patent Office, the job Einstein was doing at that time. Today It is accepted as true for everything from *decaying uranium to fission bombs to nuclear fusion in the Sun and matter antimatter annihilation.*

The mass destroyed becomes energy quantified by $E = mc^2$. The third and most profound consequence is that *energy can be used to make mass out of nothing.* It is difficult for the human mind to accept that this is true.

Imagine two rapidly moving billiard balls smashing into each other; after the event you would not expect to see anything other than two billiard balls.

In the quantum world however it works out differently. If the energy level of smashing fundamental particles in a particle accelerator is sufficiently high as in the Large Hadron Collider in Europe, you can generate material particles like a *photon,* an *electron,* or a new matter antimatter pair like an electron (particle) and *positron* (antiparticle). This phenomenon is not visible in day to day events but becomes possible when enormous energy is generated as in the Light Hadron Collider in Europe.

Particle accelerators constantly search for and find new, unstable, high energy particles like the *Higgs* boson (*God particle* believed to have triggered the Big Bang that created our universe) or the *top quark* (heaviest particle that strongly interacts with the Higgs boson)

In all such cases the mass you get comes from the energy generated as per the formula $m = E/c^2$.

There is a weird twist to this statement. In the quantum world, due to *wave particle duality* and its consequent *Heisenberg uncertainty,* there is an inherent *unknowability* to the created mass at a predetermined moment.

It is difficult for the layperson to wrap his head around this counter intuitive reality in the quantum world, but it is as true as night follows the day. By contrast, in our day to day macro world the mass and velocity of a moving object, like an earth satellite can be precisely determined continuously throughout its flight.

Mass energy *equivalence* led Einstein to his greatest achievement, the *General Theory of Relativity*. It took him 10 long years to bridge the gap between the Theories of Special and General Relativity. For a better understanding of this abstract topic by the layperson, we propose the following thought experiment.

Imagine a particle/antiparticle pair moving rapidly, as though they had fallen from outer space immediately after the Big Bang. They annihilate each other close to the surface of the Earth. The created photons would now have extra energy, not just from $E = mc^2$, but the additional kinetic energy they gained by their fall.

If as per classical physics this energy is to be *conserved* we have to conclude that *gravitational redshift* is real i.e. the more distant an astronomical object is from us, the *faster* it is moving *away* from us.

Newton's laws of gravitation had no way of explaining this, but in Einstein's General Relativity, the curvature of space means that falling into a gravitational field causes a gain in energy, and climbing out of it an energy loss. The relationship between mass and energy for a moving object, is therefore not just $E = mc^2$ but more precisely

$E^2 = m^2c^4 + p^2c^2$ (where $p$ stands for momentum).

Ahead of Newton, Einstein was able to realize that only by generalizing moving astronomical objects to include energy, momentum, and gravity could we truly describe the universe.

The above proviso notwithstanding, Einstein's equation, $E = mc^2$, is a triumph of the power, simplicity and profound implications of basic physics that

*Matter has an inherent amount of energy within it, mass can be converted to pure energy, and energy can be used to create massive objects that did not previously exist.*

This thinking has enabled us to discover fundamental particles, to invent nuclear power for peaceful and destructive purposes and reveal an aspect of gravity that describes how every object in the universe interacts with every other object. We next focus on the related topic of,

*The difference between the speed of light and the speed at which the universe is expanding.*

Simply stated the difference is that while the universe is expanding and accelerating ever since the Big Bang at mind boggling speeds, the speed of light - 300,000 kms per second remains constant.

Much of the universe's expansion occurred a fraction of a second after the Big Bang, during *inflation* and the initial positions of all the matter in the universe are imprinted on its afterglow.

Einstein's Special Theory of Relativity that nominates an upper limit to the speed of light relates to *what one person sees when he looks at another person moving at constant speed relative to the two of them only.* With the General Theory of Relativity published in 1915 Einstein extended special relativity, relating what one person sees to the general case when they look at another person accelerating relative to them. The theory fortuitously turned out to also be a theory of *gravity* which can be applied to the biggest gravitating system of all - the *universe.* And, when it is, it reveals *space time* to be a kind of background to which the *galaxies* are effectively nailed. That *backdrop* not being a material thing can expand at any rate it likes. And it certainly did just that.

That was during the epoch of inflation, during the first split - second of the universe's existence, when the *expansion of the universe* occurred at a rate that was many fold faster than the speed of light.

Gravity had a big role to play. It is the weakest force in our day to day world and yet the *strongest* force in the universe. Winston Churchill had facetiously described gravity as - a *riddle, wrapped in a mystery, inside an enigma.*

This means that on an astronomical scale, objects are getting farther and farther away from one another and the ultimate prospect of a *vision limit* for sentient beings on Earth would occur when our parent Milky Way Galaxy merges with the adjacent Andromeda galaxy.

Why?

Because light from the remaining estimated 125 billion galaxies in the universe, will not be able to move fast enough i.e. *faster than the speed of light* to compensate for the expansion of the universe. In the night sky we will be able to *see stars from a single merged galaxy surrounded by an ocean of darkness. Nothing else will be visible.*

Light exists in the universe but it does *not constitute* the universe. The expansion is about the universe, and not the bodies in it. The Earth is not getting bigger, our solar system is not getting bigger, only our galaxy size will change when it collides and merges with the Andromeda galaxy, about *4.5 billion years* from now.

*Space….the Universe ….can expand faster than the speed of light.*

This is tacitly implied in the Big Bang Theory, the best explanation we have for the commencement of our universe 13.8 billion years ago and its perpetual expansion thereafter faster than the speed of light. Let us probe the expansion of space further. Unlike the routine scientific research conducted in universities and allied institutions, space studies are more in the realm of theory and speculation rather than based on fact. The current rate of expansion of space is estimated to be 73.3 kms per second per mega parsec. The surprise element is - *it is the same everywhere.*

From the angle of the expansion of the universe, it is as if the Big Bang happened *everywhere* at *the same time*. This homogeneity contradicts our most accepted model of the Big Bang and creation of the universe. We do not as yet have an answer to this dilemma.

What is a *mega parsec?*

In astronomy it is a unit of measure of distance i.e.one million parsecs is equal to 3.26 light years.

Therefore, when we speak of kms per second per mega parsec we mean a measure of speed since you have units of both distance and time. *The Hubble Constant* is defined as a km per second per mega parsec. The term is frequently used when discussing space related topics. The dividing line between astronomy and space studies on the one hand and religion and folklore on the other often gets blurred. Over time this leads to half truths and myths.

Has your kid innocently asked you,

*Dad, if you took a spaceship far into the universe would you come back to where you started?*

Had the question related to Earth, you would unhesitatingly reply in the affirmative but in respect of space you would be compelled to admit that you do not know the answer. In fact *nobody knows for sure.*

Why?

Even if the universe is spatially closed like the Earth, it is expanding so rapidly that even at speeds approaching the speed of light, your spaceship would never come back to the point from where it started. Think of the universe as the surface of a balloon blowing up. Stars and galaxies *dot* its surface. The farther apart two galaxies are, the faster they recede from one another. As you travel away from your home galaxy at near speed of light, space in front of you is actually receding from you *faster* than you can reach it. That means you can *never* reach galaxies beyond a certain distance from you. The answer to your kid's innocent question is mind bogglingly complex.

In this context, the *Hubble sphere,* an imaginary sphere in the heavens beyond which galaxies are receding away from us faster than the speed of light is relevant and interesting. Modern cosmology enables us to explore this unknown esoteric region through *absorption and emission spectroscopy* experiments. They indicate that almost all galaxies are receding *away* from us because their light is redder than what it should be. In this context, a frequently asked question is whether objects receding faster than the speed of light are visible?

*Surprisingly, they are visible.*

Unlike the *event horizon* on the periphery of a black hole that can be crossed with a one way ticket *only,* objects outside the Hubble sphere can influence those within it and vice versa.

Summarizing, there are four important regions in our universe:

- The Hubble sphere

- The *particle horizon* i.e. the distance light emerging from the Big Bang moment could have reached by the present day.

- The *event horizon* i.e the region from which light can reach us from the Big Bang to infinite time.

- The past *light cone* that contains every point in the universe from which light could possibly have reached us from the Big Bang to the present day.

This light cone is about *46 billion light years* wide and constitutes the observable universe. With this backdrop, it is interesting to probe popular myths related to the cosmos.

*Myth No. 1: Objects cannot recede faster than the speed of light.*

Special Relativity that specified the constancy of the speed of light, applied only to *relative motion of inertial frames.*

In other words, if two observers are in constant relative motion and experience no other force, then their measurement of the speed of relative motion *cannot* exceed the speed of light. This gives rise to the phenomenon of *time dilation.*

In an expanding universe, however, we are talking about *non inertial frames* where observers are not moving in space, rather *space is itself expanding.*

Thus, they perceive themselves getting farther apart because distance itself is changing. While Special Relativity *cannot* explain this phenomenon of redshift of *space itself* we can derive partial comfort from General Relativity.

Hubble initially failed to realize this shortcoming in Special Relativity, and got his velocity calculations wrong.

For many years, until the end of the 20th century, the difference between Special and General Relativity red shifts had not mattered because observations were not deep enough into space. The redshifts are similar for objects that are closer but diverge *significantly* for objects that are tens of billions of light years away.

*Myth # 2: Superluminal expansion occurs only in the Inflation Theory.*

You will recall the *Inflation Theory* that in the early stages of our universe, expansion was at a much faster rate than at present. This led to the use of loose terminology like *superluminal* describing expansion faster than the speed of light. The difference between our early universe and the present day is that the Hubble Constant was much larger then. If the *exponential expansion* of the early universe had continued to the end of time the *Hubble sphere would eventually have coincided with the event horizon.* In such a situation, you would not have been able to *observe* objects outside the Hubble sphere for a very long time after the Big Bang. If inflation caused all distances in the observable universe to scale down to the *Planck length* to expand faster than the speed of light, the use of the expression *superluminal expansion* is justifiable.

It would then logically mean that the size of the Hubble sphere and Planck length would be the same which is obviously not true. Hence *faster than light expansion* describes reality more accurately than *superluminal expansion*.

*Myth # 3*

*Galaxies receding faster than the speed of light exist but we cannot see them.*

Occasionally there are reports in science journals, that even though galaxies can recede from us faster than light we cannot *see* them. This is not true.

The speed of photons heading towards us is *not* the speed of light. It is the recession velocity minus the speed of light. This weakness in Newtonian thinking is because while light propagates at its stated speed in a vacuum, when space itself is expanding it is like *light moving on a treadmill*.

Like a rocket traveling to the edge of the universe to check whether it has come back to its starting point?

The rocket could be moving at the speed of light but the expansion of the universe would neutralize the effect.

Following this line of thought, it would seem that there is no way we can *see* this light.

Strangely, this is *not* true.

The Hubble sphere also recedes. As long as this recession faster than the photons immediately outside it is in place, we will be able to *see* those distant galaxies.

The slowing rate of expansion with distance allows the Hubble sphere to *overtake* photons just outside it.

If this were not the case, the Hubble sphere would coincide with our event horizon.

Proof?

*Hundreds of galaxies with redshifts above 1.46 that have been observed.*

The highest redshift observed so far is *GN - z11 in Ursa Major at 11.1* and at least five galaxies with redshifts above 8 have been *observed*. Our particle horizon determines how far we can see, *not* the Hubble sphere. Theoretically, we should be able to see formations all the way back to the Cosmic Microwave Background (CMB) at 1100 redshift.

Points where the CMB was emitted are currently receding at 3.2 times $c$ the speed of light. When emitted their speed was a whopping 58 times $c$

*Myth # 4: We do not know if general relativity is a good description of cosmic redshift.*

This may seem logical from the angle that we have no way of corroborating velocities and distances of remote galaxies other than the redshift model based on general relativity. This assumption is false.

In *Type Ia Supernovae* we have a good way to measure time dilation. These supernovae provide *light curves* or timed events with 60% - 80% maximum intensity. For us on Earth, we find that these supernovae events take longer than they would for a local observer. Measurements of this time dilation are consistent with the general relativistic picture but *not* with special relativity.

Another test with these supernovae is their magnitude in turn related to distance and therefore a good method of interpreting the velocity of the receding galaxy. On the other hand if you interpret the magnitude using special relativity you end up far away from the observed redshift. General Relativity *provides a very good match*.

## Conclusion:

Increasingly, we are learning that superluminal speeds are a regular feature of General Relativity.

While Special Relativity is a poor model of cosmic redshift, there are several realistic General Relativity models that predict slightly different velocities for an observed redshift. Be that as it may, there are other observables that corroborate our interpretation of velocity and distance. *Distance* in a cosmic context is subject to more than one interpretation. Reverting to your kid's question on finally returning to the starting point of a grandiose space odyssey we may never know the answer as a simple *yeah* or *nay*.

For a positive result, humans would have to learn to warp space, create rockets that can travel faster than the speed of light and thereby explore beyond *what we can ever see.*

Before we conclude this chapter on the *Speed of light* we shall briefly review a notional fundamental particle called the *Tachyon or faster than light particle.*

Tachyons are purely theoretical - there

Is no observational evidence that they exist. Special Relativity prohibits travel through space at speeds exceeding light speed.

In 1967 purists pointed out that Special Relativity did not actually prohibit this but only prohibited an object that is initially moving at less than light speed from accelerating to speeds exceeding the speed of light. This logic got extended to the theory that there could be objects called tachyons or *fast ones* that *always* move faster than the speed of light. As they lose energy they are predicted to move faster and faster *without limit.* Such objects - if they are to have observable energy are required to have *imaginary mass* i.e.instead of having a mass of 1 kg they would have a mass of $1i$ kg.

Albert Einstein in a lighter vein mentioned in 1907 that if he shot someone with a *tachyonic gun* then a third person moving sufficiently fast would observe the death event even *before* he had fired the shot. This violates the law of *cause and effect* a vital part of the structure of space time.

The current status of tachyons is that *they are not forbidden by the rules of physics* but this does not mean that they actually exist. The reasons for such a conclusion is that they would be required to have imaginary mass and would violate the law of *causality.*

Their status is of *theoretical curios* not to be taken seriously.

# Thirty Five

# Time Travel & Worm Holes

*Time travel used to be thought of as just science fiction, but Einstein's general theory of relativity allows for the possibility that we could **warp space - time** so much that you could go off in a rocket and return **before** you set out.*

**Stephen Hawking - English theoretical physicist (1942 - 2018)**

This essay takes us to the hazy dividing line that separates reality from science fiction.

Albert Einstein's *Theory of General Relativity* introduced a stream of new thoughts to the Classical understanding of *space* and *time*. It was the starting point of the revelation of some of nature's deepest mysteries. One was the presence of *black holes* and the related phenomenon of *worm holes*.

Worm holes were first theorized in 1916 by Austrian physicist *Ludwig Flamm* as *bridges connecting different points in spacetime.*

*In theory they provided short cuts for space travelers through time reversal of a black hole. Entrance to the black hole could be through a space time pathway.*

In 1935, the bridges through space time idea to reduce travel time was reiterated by Einstein and Rosen leading to the appellation *Einstein Rosen bridges* or wormholes. Stephen Hsu, Professor of theoretical physics at the University of Oregon, USA described the development as hypothetical saying,

*No one thinks we are on the verge of finding a wormhole anytime soon.*

Be that as it may, speculation on this fascinating topic continues unabated.

Wormholes a result of warping of spacetime, have been theorized to consist of two mouths with a connecting throat.

The mouths are likely to be spheroidal. (*Journal of High Energy Physics 2020*)

The throat could be a straight or winding stretch.

Einstein's Theory of General Relativity math *predicts* the existence of wormholes but none have as yet been discovered. Some solutions to the general relativity equations provide clues that the wormhole mouth could be a black hole, but our understanding of a black hole originating out of a dying star does *not* permit this. In the realm of science fiction, the problems of travelling through wormholes could be summarized as follows:

- Problem of *size:*

  Primordial wormholes are predicted to exist only in microscopic levels, about $10^{-33}$ centimetres; however as the universe expands, the possibility that some could get stretched to larger sizes cannot be ruled out.

- Problem of *stability:*

  The Einstein Rosen wormholes as predicted would be useless as they would collapse too quickly. Theoretical physicist Hsu is of the view that an exotic type of matter would be required to resist such a collapse. There is no evidence as yet of the existence of such exotic material or possibility of its creation. Although still in the realm of imagination, several recent studies have suggested intriguing ways to move forward. Black holes and wormholes are special types of solutions to Einstein's field equations, the mathematical foundation on which the Theory of General Relativity is structured. The basic principle is that spacetime is strongly bent by gravity as in a black hole where the fabric becomes so curved that not even light can escape from it. The Theory of General Relativity allows spacetime to be freely stretched and bent that can result in a plethora of configurations that include wormholes.

As stated, wormholes that are sufficiently large and stable enough to resist gravity may be traversable (travel worthy) since it will try to close them.

This in turn would require a large quantity of *exotic matter* i.e. matter with negative energy. It is difficult to understand this concept since the matter we see in our daily lives has *positive energy only*. As of today, the laws of quantum mechanics make it theoretically possible to create exotic matter or negative energy but in *very* small quantity and for *very* short period of time only. At the same time there is no evidence that the quantity of exotic matter *cannot* be increased in the future. There is also the possibility that quantum mechanics scientists discovering a new set of equations that enable time travel *without* exotic matter.

There is another significant problem of an academic nature. Time travel seems to contradict logic in the form of *time travel paradoxes* like the *consistency paradox*.

What is a consistency paradox?

It describes an event that changes the *past* but the change itself prevents the event from happening. Consider that you enter a time machine and go back in time by 1 hour and destroy the machine. Since the machine is destroyed it would be impossible for you to use it an hour later. But if you cannot use the time machine then the question of going back in time by an hour does not arise. The contradiction that arises is that the machine cannot be destroyed and *not* destroyed simultaneously and *ipso facto* you cannot change the past. For this reason would be time travellers are cautioned *not* to attempt to change the past. In physics a paradox cannot happen because of the inconsistency in the theory that upholds it. It can *never* actually happen.

Stephen Hawking formalized this truism in what he called the *chronology protection conjecture*.

Attempts to resolve time travel paradoxes were made by theoretical physicist *Igor Dmitriyevich Novikov* through the *self consistency conjecture* that states that you can travel into the past but cannot change it. So, is this the final nail in the coffin for aspiring time travellers?

Not quite if you allow *multiple histories*. Parallel time lines can resolve paradoxes that Novikov's conjecture cannot. When you exit the time machine, you exit into a different timeline. Here you can do whatever you want including the destruction of the time machine without changing anything in the first timeline!

Everett's *many worlds interpretation* where one history can split into *multiple histories* suggests that it is within the realm of possibility. Researchers are working on a concrete theory of time travel with multiple histories that is compatible with General relativity and quantum mechanics.

Be that as it may, at a practical level how do we go about finding a worm hole in the prodigious vastness of space?

We do not have a precise answer. Scientists look for them at the centre of bright galaxies and telltale spectacular display of high frequency gamma rays.

In this method of detection, however there is the possibility of mistaking them for black holes. The only difference from an observation angle is that black holes produce less gamma rays and eject them as a jet whereas radiation from a worm hole would be in the form of a gigantic sphere. To permit human travel, temperatures need to be bearable and therefore the worm hole needs to be a minimum distance away from the galactic centre inferno. Discovering a wormhole requires painstaking tracking of the orbits of stars close to a black hole.

There is a school of thought that the supermassive black hole at the heart of the Milky Way Galaxy may have been caused by a wormhole. This is a field of unending speculation. Unlike treacherous black holes that trap anything that enters them including light, wormholes offer the exciting prospect of travel to unexplored regions faster than the speed of light!

The universe is huge. Travelling at the speed of light to *Proxima Centauri* the nearest star would take more than four years. Venturing to the other side of the galaxy?

*More than 100,000 years.*

So what are the choices before the intrepid space traveller?

He could be inspired by the film *Interstellar* by Christopher Nolan and take the plunge or give it up as a lost cause. But wait. Let us try to intensify our intellectual probe into this abstract topic. Imagine the universe as a two dimensional sheet. Poke two holes and curve the sheet around them to form two funnels. Stitch the ends of the funnels together to create a *wormhole like tube.* If you can stretch your imagination a trifle further, it would clarify how a journey across the two funnels would be *much shorter* than through the vast unexplored tracts of space and time.

This thinking encouraged the idea of the Einstein Rosen bridges as a *shortcut* to wormholes enabling time travel. Similar thinking inspired *Carl Sagan* to write his novel *Contact* with the help of physicist *Kip Thorne* to enhance the credibility factor. There are many counter intuitive realities that have to be overcome to forge ahead e.g. if exotic matter or negative energy was applied on Earth you would need to swing your racket *away* and not *towards* a negative energy tennis ball to hit it!

We could go on and on on this Knowledge Frontier Area. The last word has not yet been stated.

Stephen Hawking makes the point in his last book *Brief Answers To The Big Questions* that wormholes may well turn out to be the time machines visualized by futuristic writers like H.G.Wells that enable humans to travel backwards.

In the brave new world of quantum physics, particles can pop out of empty space and disappear a moment later. This begs the question, if particles can be created, why not wormholes?

Physicists believe primordial wormholes may have been formed in the early universe from a *foam of quantum particles popping in and out of existence.* It is possible some of them are around even in the present day. Recent experiments on *quantum teleportation* have shown that disembodied transfer of quantum information from one location to another bears a mysterious similarity to two black holes connected through a wormhole.

Such experiments appear to resolve the *quantum information paradox* that suggests physical information could permanently disappear in a black hole. Interestingly, they could also reveal a hidden connection between the incompatible theories of quantum theory and gravity. Wormholes are relevant to both and could turn out to be instrumental in their reconciliation leading to the construction of a *Theory Of Everything* (TOE)

The fact that wormholes play a role in such developments is unlikely to go unnoticed. We may not have yet seen wormholes, but they could certainly be out there. They may even help us understand cosmic mysteries like *multiverses* that beg the question whether *our universe is the only one that exists?*

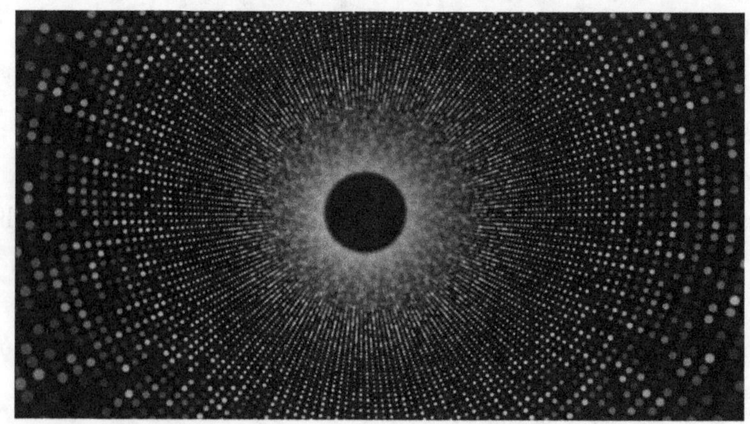

**Illustration 21:** Wormholes- Shutterstock _ 2131825391

# Thirty Six

# Quasars, Blazars & Pulsars

*New ways of seeing can disclose new things. The radio telescope revealed quasars and pulsars, and the scanning electron microscope showed the whiskers of the dust mite. But turn the question around - Do new things make for new ways of seeing?*

**William Least Heat - Moon American Travel Writer (1939 - )**

Pulsars, Blazars, and Quasars sound like cousins; however they are astronomical bodies with distinct identities. A pulsar is a star, blazars and quasars are galaxies.

Pulsars are highly magnetized rotating *neutron* stars (much smaller than our sun) while blazars and quasars are gigantic, extremely powerful and distant active galactic nuclei. Summarizing the major differences:

Quasars and blazars are *much bigger and brighter* than pulsars. They (Q &B) are *more distant* from Earth than pulsars. It is no exaggeration to state that a chalk and cheese difference exists between Q &B on the one hand and pulsars on the other. Most striking is in respect of size. A typical pulsar can be described as a small star a bare *20 kms in diameter* approx the distance from Kempegowda International Airport, Bengaluru to the Cantonment train station in the heart of the city. By cosmic standards this size is minuscule.

What makes a pulsar special is that it has a beam of light that swings around like a lighthouse causing the star to appear to *pulse*. A pulsar is a neutron star that is *spinning in a special way.* The reader needs to get used to the *spinning* concept because everything in our universe not only moves but also spins*!*

A good example is planet earth that spins on a tilted axis. As explained earlier in our narrative, a single rotation in an average time span of 24 hours gives us day and night. The second movement of revolution in an elliptical orbit around the sun in 365 days establishes weather patterns in the various regions of the planet.

The layperson is unlikely to be familiar with another aspect of its spinning property - Earth has 2 *North poles* and 2 *South poles.* One set represents the tips of the axes the earth spins on, the North Pole [(Arctic region) and the South Pole (Antarctic region) visible on maps.

The other set is less obvious. It is caused by the earth's magnetic field with a north side and a south side. These two sets of poles are *slightly misaligned.* (The Earth's magnetic field can be described as a magnetic dipole with the magnetic south pole located near the Earth's geographic north pole, and the magnetic north pole near the geographic south pole. This is why the compass used in navigation points to the geographic north pole since it gets attracted by the magnetic south pole.

The same logic explains the proximity of the geographic south pole to the magnetic north pole ]

Both magnetic poles are *misaligned along the Earth's rotational axis by about 11.3 degrees.*

Reverting to pulsars, the poles are significantly misaligned, more than on earth. The swinging lighthouse of a pulsar appears in the night sky as a *super bright light beaming out from the magnetic poles.* Light could include other parts of the electromagnetic spectrum like *X Rays, Gamma Rays* that are invisible to the human eye.

What are quasars and blazars?

The comments that follow on quasars apply equally to blazars unless stated otherwise. A quasar is the middle section of an active galaxy encompassing a number of stars - it is short for *quasi stellar radio source* also called QSO short for *quasi stellar object.*

Quasars were discovered as recently as 1950. Initially they were objects of mystery as their spectrum did not conform to any known star types and hence described as *quasi stellar!*

Space scientists realized that the weird spectra of quasars was because of their mind boggling distance from earth.

*They were further away than anything that had ever been seen.*

They could be up to 13 billion light years away, a record in astronomical distance and brightness since they are visible despite the humongous distance that separates them from earth. Scientists describe them as a *halo of high energy matter around a black hole at the centre of a baby galaxy.* An interesting question is - Are our *Milky Way Galaxy* and neighbour *Andromeda Galaxy* home to quasars?

The query is incomplete without a time frame considering the long history of our ever expanding universe over gigantic swathes of time. We can state that at the present moment the answer is *no* because had there been quasars in the Milky Way and Andromeda Galaxies, their blinding light would have blanked out trillions of stars, planets and nebulae presently visible in the night sky.

It is reasonable to presume that in the early galactic history of our universe nearer the time of birth of the Milky and Andromeda Galaxies, when they were rapidly moving away from us, *quasars did exist* in them. In other words they commenced their life sketches as quasars.

Be that as it may, it is now known that quasars are distant extremely luminous bodies whose light on reaching Earth is *redshifted due to the metric expansion of space.* (Hubble's law states that remotely distant light sources are redshifted proportional to their distance from Earth)

Quasars inhabit the centres of active galaxies and are among the most luminous, powerful, and energetic bodies in the universe emitting up to a *thousand times the energy output of the present day Milky Way Galaxy that contains 200 - 400 billion stars.*

The energy is in the form of radiation across the electromagnetic spectrum, normally peaking in the ultraviolet optical band. Some quasars are strong sources of radio emission and gamma rays.

With high resolution imaging from ground based telescopes and the Hubble Space Telescope, the *host galaxies* surrounding the quasars have been detected in a few cases.

(Galaxies are too *dim* to be seen against the glare of the quasar. Most quasars cannot be seen with small telescopes)

Quasars are in many cases powered by accretion of material into supermassive black holes in the nuclei of distant galaxies. We know that radiation including light cannot escape from within the event horizon of a black hole. The energy produced by a quasar is generated outside the black hole, by gravitational stresses and immense friction within the material nearest to the black hole as it orbits and falls inward.

The blinding luminosity of quasars arises from the accretion discs of central supermassive black holes that can convert 6% - 32% of the mass of an object to energy.

(*The immensity of the energy generated can be assessed by the relatively low 0.7% mass energy conversion happening in the savage fusion reaction within our Sun*)

Central masses of 105 -109 solar masses have been measured in quasars.

As per present thinking, galaxies similar to the MWG and Andromeda that do not have quasar activity contain similar supermassive black holes in their nuclei.

Only a small fraction have sufficient matter at their centre to become active and power radiation that provides them visibility as quasars. This explains why quasars were common in the early universe - energy production ceased when the supermassive black hole had consumed all the gas and dust near it. As stated earlier, we can reasonably conclude that most galaxies including our MWG and adjacent Andromeda had in the early stages of their evolution gone through a quasar active phase. These galaxies subsequently became quiescent because of lack of supply of matter into their central black holes to generate radiation. It is predicted that a quasar could form when Andromeda collides with our MWG, 3 - 5 billion years from now.

In 1980 unified models were developed in which quasars were classified as a kind of active galaxy. The present day consensus is that they differ from *blazars* and radio galaxies only because of the viewing angle:

*If the angle of the stream is slightly towards us, then it is a quasar and if the stream is angled directly towards us it is a blazar.*

At the cost of repetition, a brief summary is given below:

- Quasars and blazars are not stars, they are a type of galaxy.

- A quasar is an active galactic nucleus that emits a powerful stream of particles from its centre.

- The difference between a quasar and a blazar is *only* on how it is angled towards us.

- Our Milky Way and adjacent Andromeda Galaxies had been a quasars in the distant past but are presently quiescent.

- The nearest quasar to Earth is *Markarian 231* in the constellation of *Ursa Major.*

- The first quasar was discovered on 5th August 1962.

- More than a *million* quasars have been found. The closest to Earth is *600 million light years* away.

- Quasar discovery surveys show that quasar activity was more common in the distant past, the peak epoch estimated to be *10 billion years ago.*

- Concentrations of multiple, gravitationally attracted quasar groups constitute some of the *largest known structures in our universe.*

Quasars fall in the category of *the most mysterious celestial bodies in the universe!*

# Thirty Seven

# Quantum Entanglement

*Quantum physics teaches us that we can simultaneously exist in many places, under certain conditions.*

**Dr.Amit Ray (1960 - ) Indian author, philosopher and spiritual master.**

Quantum computers and quantum cryptography are in the news these days. Their functioning depends on *entanglement,* a property of quantum physics that make such magical devices possible. Aside from this, Quantum Entanglement (Q.E.) can be used to make more accurate measurements of *gravitational waves* (ripples in space - time) and to better understand the properties of *exotic materials* (like plastics, semiconductors and ceramics)

It also subtly shows up in other places e.g. when atoms bump into each other and get entangled affecting the accuracy of even *atomic clocks.*

What is an atomic clock?

They are the most accurate clocks in the world e.g. the *strontium atomic clock* is accurate to within 1/15,000,000,000 of a second per year. Put another way, it would not have gained or lost a second if it had started at the dawn of creation of our universe, the Big Bang.

What exactly is *entanglement?*

For a better understanding we need to revisit three laws from the Classical and Quantum eras of theoretical physics:

## Conservation of Energy:

The Law of Conservation of Energy states that the total amount of energy in an isolated system remains *fixed.* (although the form is convertible - from electrical to mechanical to heat etc).

This law governs the working of all machines, whether steam engines or electric cars. Conservation laws are like accounting statements - you can exchange bits of energy around, but the total amount has to stay the same.

## Conservation of Momentum:

The Law of Conservation of Momentum is exemplified in respect of two ice skaters with different masses pushing off from each other, the lighter one moving away faster than the heavier (momentum equals mass into velocity expressed as $p = mv$).

## Conservation of Angular Momentum:

The Law of Conservation of Angular Momentum is clarified by the whirling skater drawing her arms closer together in order to spin faster. These Laws of Conservation of Mass, Energy and Angular Momentum have been experimentally verified over colossal size ranges - from black holes in distant galaxies down to the tiniest spinning electrons. An important concept related to entanglement is:

## Quantum superposition

Imagine you are trekking in unknown territory and the trail leads to a fork with a distinct left or right option. The left is cold but promises picturesque viewing. The right is bright and sunny but with a steep climb. You opt to turn right. In our day to day world, without further ado you would turn right and proceed with your trek.

*In the quantum world in contrast, under certain conditions you could simultaneously choose both options.*

How?

For systems described by Quantum Mechanics (Q.M.) the rules differ from Classical Mechanics (C.M.) They are counter intuitive and more intriguing. An electron for example, governed by the laws of Q.M. can be in a quantum state where it spins *clockwise* and shift to another state where it spins *anti clockwise* or even be in a third state where the spin is *clockwise + anti clockwise.* That is not all. The quantum states can be added together or subtracted from each other *not* as mathematical scalars but as vectors.

Such a combination of quantum states is referred to as a *super position.*

In layman's language it means *being in two positions at the same time.*

This intriguing phenomenon was first demonstrated by British researcher Thomas Young as far back as 1801.

Through his *double slit experiment* he proved the *wave particle duality of light.*

The unpredictability of the end result wave or particle status *till observed* is what makes Q.M. distinct from the definiteness associated with C.M. experiments. It is interesting to observe the effect of combining the Conservation Laws of C.M and Q.M.

Imagine a pair of atoms having 100 units of energy. The first atom is with you and has 40 units. To conform to the Law of Conservation of Energy, your friend with the second atom would carry 60 units of energy. Within the atom energy rests in the *electron orbital.* An unusual property of electrons is that they could occur in *pairs.* The working of electron pairs is intriguing. They bond irrespective of the distance that separates them. The bond is valid in all respects whether the second electron is a microscopic distance from the first or millions of miles away in another galaxy. Einstein was initially skeptical of this claim since he believed that *nothing* in the universe could travel faster than the speed of light. He spoke derisively of the phenomenon as *spooky action at a distance.* However subsequent events forced him to soften his stand. In 1949, *Chien - Shiung Wu* and *I.Shaknov* confirmed the validity of *spooky action* in the lab. The result specifically proved the quantum *correlations* of a pair of photons. This generated introspection within the scientific community. Is the reported lab result a fluke event or *are all atoms in the universe entangled?*

Modern cosmology seems to indicate that most particles in the visible universe exhibit a high degree of entanglement.

An interesting feature is the effect of Q.E. on humans. A study suggests that unexplainable phenomena like *faith healing* could be the correlation between the atoms of the healer and the healed that get awakened under certain conditions. The wide implications of these findings split the scientific community into two camps - The conservatives who stick to the principles of C.M. and the progressive keen to expand the knowledge frontier open up to Q.M. The troubled waters were stilled by Irish physicist *John Stuart Bell* (1928 -1990) through his *Bell Inequality Theorem*. Bell's clarification was timely and decisive coming as it did in the face of stiff opposition from the conservative camp led by *Einstein, Podolsky and Rosen*. They formulated the *EPR Paradox* purporting that Q.M. theories were incomplete and contained unexplored *hidden variables*.

Bell's Theorem on the other hand devised a means of testing whether particles connected through quantum entanglement could communicate information faster than the speed of light.

The theorem stated that no theory of local hidden variables could account for all the predictions of quantum mechanics. He created Bell inequalities, that are violated in quantum mechanics systems. This led to the inference that the idea of local hidden variables was false. In 1970 the scientific community finally accepted the Bell notion that no physical effect can move faster than the speed of light had been *effectively falsified*.

We need to understand the significance of the phenomenon of *correlation* in the quantum world as opposed to *causation* i.e.cause and effect in our day to day lives. The mysterious phenomenon of correlation in theoretical physics that leads to *Quantum Entanglement* is unexplainable by the Laws of Classical Mechanics. For a better understanding of Q.E.by the reader, we give below a few explanatory comments:

## Quantum state:

Various states of a microscopic system that includes an electron. It describes its charge, spin, and energy level that can be measured simultaneously.

## Quantum spin:

The spins of subatomic particles like electrons resemble the spins of macroscopic bodies like planets. Electron spin refers to its angular momentum. The spin *quantum number* provides information on the electron's unique quantum state. All particles have an intrinsic spin but visualizing it in the macro world is extremely difficult, almost impossible.

## Quantum number:

Value that describes the energy level within an atom. An electron has quantum numbers to describe its state like *principal, azimuthal,* and *magnetic.*

We refrain from probing this abstract topic further since it does not contribute to the message this essay seeks to convey. In 1970 the scientific community after extensive debate accepted the tenets of the Bell theorem as the correct interpretation of quantum entanglement. The important takeaway was that *the classical concept of locality* was not necessarily applicable to the quantum world.

## Temporal Entanglement:

In the narrative so far the focus has been on space and positioning. It can be extended to the *time* dimension also.

The temporal non locality experiment was conducted in 2013 by physicists in the University of Jerusalem, and here, as they say, is where *ye olde plotte thickeneth.* The foray into time entanglement is indeed a moment of excitement for the theoretical physicist.

I am digressing at this stage from the intellectual heights of Q.E. to a more mundane level.

Delving into English literature I recall the words of Sherlock Holmes to Watson at the beginning of *The Adventure of the Abbey Grange:*

*Come, Watson, come. The game is afoot. Not a word. Into your clothes and come.* Arthur Conan Doyle, creator of Sherlock Holmes had borrowed the expression first from Shakespeare's *Henry the Fourth* and yet again from *King Henry the Fifth.* In both cases the context is *intensely time related* that the hunted animal had been sighted and the chase had begun. Temporal entitlement is similarly exciting and *intensely time related.* A team of physicists at the Hebrew University of Jerusalem reported that they had successfully entangled photons that *never coexisted.* Previous experiments involving a technique called *entanglement swapping* had already shown quantum correlations across time, by delaying measurement of one of the coexisting entangled particles. *Eli Megidish* and team showed for the first time that the entanglement phenomenon could happen with photons whose lifespans did *not overlap.*

How?

They conducted experiments involving *Five Distinct Steps or Phases.*

Details of each step would lengthen our narrative without contributing to the message of this essay and are hence being excluded. Instead, I am summarizing the conclusions arrived at by the Eli Megidish team:

- Existence of quantum correlations between *temporally non local photons* i.e.between two quantum systems that never coexisted.

- The significance in layperson's language is that the *polarity of starlight from the distant past say five times the Earth's lifetime could influence the polarity of visible starlight during your next viewing season few months from now.*

- Even more bizarre, your next set of measurements relating to the polarity of starlight somehow dictates the polarity of photons *more than 15 billion years old.*

- In both forward and backward directions, quantum correlations span the *causal void* between the death of one photon and the birth of another. *Coexistence* is not a prerequisite.

Shifting gears to the spiritual plane, we recall Chapter 8: 58 of the Gospel of St.John where Jesus says,

*Verily verily I say unto you, before Abraham was I am.*

It is indicative of a possible quantum relationship to *consciousness and the flow of time.*

*Perhaps the replacement of yesterday, today and tomorrow* in the words of the celebrated hypnotist Paul Goldin by *the everlasting now.* We are aware of time moving fast or slow depending on circumstances. What is now being theorized is a scientific backing for the extreme position of time coming to a standstill and events being explained through *a flow of human consciousness.*

A movement in this direction was made in the Special Theory of Relativity, when the Classical concepts of *absolute time* and *simultaneity of events* were toppled and the concept of a *single time keeper* for the universe demolished. To maintain the tempo of this high octane topic, we need to permit a dose of *willing suspension of disbelief.*

How?

For starters, we pay serious attention to *personhood and individual and group consciousness flows to explain events rather than the unreliable classical concept of time flow.*

Physicists are coming around to realize that that this is a possible way out of the quantum jungle!

**Illustration 22:** Quantum Entanglement - Shutterstock _1984049708

# Thirty Eight

# Interpretation Of Dreams

*I believe in everything until it is disproved. So I believe in fairies, the myths, dragons. It all exists, even if it is in your mind. Who is to say that dreams and nightmares are not as real as the here and now?*

**John Lennon (1940 - ) English singer, songwriter, musician and peace activist.**

The subject of *dreams* has fascinated humankind since time immemorial. Even today, in the 21$^{st}$ century it is a topic of intense debate across a wide spectrum of society in many parts of the world.

*Our Universe An Unending Mystery,* Chapter 34 provides thought provoking ideas on this commonplace topic experienced by all but understood by few. We strive to enlighten the reader further. A question often asked is:

*Are dreams nothing but our own image of a parallel universe emerging from a parallel reality?*

The following thoughts are relevant to the topic, but no *unambiguous* answers are possible.

- According to researchers, dreams are caused by random nerve impulses in the synapses in our brain.

- Most dreams arise due to poor sleep patterns coupled with anxiety, sadness and excitement when awake. Memory elements in our brain tend to provide them with concrete shapes, sizes and meanings.

- Even our own image reflected in a parallel reality is at best only a *plausible* explanation. Some dreams belong to the world we live in. It seems this linkage of dreams to our existence in a parallel universe is lacking in evidence. Similarly, the theory that it could be proof of the existence of *multiple parallel universes or multiverses* seems to be on a sticky wicket.

There is the related concept of *Maladaptive Daydreaming!*

A person prone to maladaptive daydreaming spends a lot of time in creating *structured daydreams or fantasies.* Professor Eliezer of the University of Haifa, Israel first described the condition in 2002. He related it to addictive practices like internet gaming or alcohol abuse. Also following events could act as a trigger:

- A sensitive topic of conversation.

- A picture, movie, or news story.

- Prolonged internet use.

- Sensory stimuli such as unique noises and smells.

- Distinct physical experiences.

## Symptoms:

- Extremely vivid daydreams with their own characters, settings, plots, and other detailed story like features, reflecting a complex inner world.

- Daydreams triggered by real life events.

- Difficulty completing everyday tasks.

- Difficulty sleeping at night.

- Strong desire to indulge in daydreaming for extended periods.

- Making facial expressions, whispering and talking while daydreaming.

- Awareness that the internal fantasy world is different from external reality.

The affected person may simultaneously be afflicted by another disorder called *Attention Deficit Hyperactivity Disorder* (ADHD). It is different from general mind wandering as it involves *structured fantasy narratives intentionally generated.* Is Maladaptive Daydreaming diagnosable?

- There is no universal method to diagnose this condition. The doctor may look out for the following symptoms to confirm a particular diagnosis:

- Dissociation

- Obsessive Compulsive Disorder (OCD)

- Attention Deficit Hyperactivity Disorder (ADHD)

- Any other condition that can resemble or overlap with maladaptive daydreaming.

*Possible Side Effects of Maladaptive Daydreaming:*

- Ability to focus

- Ability to be productive in work and studies.

- Attention to real life relationships.

- Mental well being, due to anxiety about managing day dreams.

- Post traumatic Stress Disorder (PTSD)

- Bipolar Disorder

- Psychosis

Despite significant achievements in the treatment and management of mental disorders in the 21$^{st}$ century, *the veil of uncertainty over maladaptive daydreaming persists.*

Experts do not know with any degree of certainty why maladaptive daydreaming happens. Possible explanations are past traumas, difficulty in managing day to day challenges and other mental conditions like ADHD and OCD.

An honest summing up would be,

*We do not as yet know.*

## Dream Symbols:

We follow a psychoanalytic approach and go beyond the Freudian *sex motivation approach* to explain dream symbols. Some frequently occurring symbols:

- *Bride*

  According to popular tradition, dreaming of oneself as a bride is an evil omen forecasting death. There is a colour twist in a western social backdrop. The white of the bridal gown changes to black indicative of mourning. Practicing psychoanalysts have clarified that the death interpretation is an extreme model and in most cases it is indicative of a fundamental change in the dreamer's lifestyle, like change of location, change of job and the like. The symbol of evil is therefore not a correct interpretation as the change that occurs could be for the good, and signify a good omen.

- *Tortoise:*

  The ancient Chinese considered the tortoise as a symbol of sacredness indicative of a divine Oracle, leading to the use of tortoise shells by seers and holy men. The Freudian view is that the tortoise shell is indicative of *a narcissistic retreat to the mother's bosom.* In contrast, Jung explains the tortoise shell as being indicative of a *collective animal unconsciousness, an evolutionary floor* that the species assumes.

- *Snake:*

  The snake is the most popular sexual symbol. The bible describes it as the *seducer that led our ancestors to commit the Original Sin. It is a symbol of temptation.* Another interpretation, based on the snake shedding its skin is a symbol of *transformation* and a good symbol. The snake is also considered to be a source for favourable medicinal inputs. In the East, snakes have symbolized fertility and rebirth, guardianship, poison, medicine and vindictiveness. In some cultures they are either recognized as gods or sacred and protected by them. The snake considered an auspicious symbol in ancient China indicated a good harvest and fertility in humans. The image of the snake is often found in their ancient relics.

The ruler of *Yunnan* province was granted a gold seal in the shape of a gold snake by the *Western Han Dynasty.*

Snake is a symbol of divinity and eternity with sections of the population in India and Egypt, unlike the Hebraic bible where it is referred to as as a symbol of the devil. Researchers tell us that the snake was the first creature to reappear after the catastrophic Great Flood of the Nile in biblical times. The Egyptian Sun God *Amun* is represented as a snake devouring its own tail signifying eternity and perpetual cycles of renewal. In India serpent deities are referred to as *Nagas.* As in Egypt, in Indian mythology the snake when biting its own tail is considered as representative of eternity.

- *Earthquakes:*

  Earthquakes are often the subject of dreams and inter related to other natural disasters like fires, floods, hurricanes, cyclones and the like.

  These are mentioned in the book of Revelations of the bible under the *Apocalypse.* Dreams depicting earthquakes can be a warning of major changes in a person's life like divorce, death, emotional breakdowns and accidents.

- *Bath Tub:*

Water is a complicated dream symbol, mostly associated with the Christian ritual of baptism. It is regarded by many as a symbol that washes away past sins. It might also be indicative of the opposite aggressive emotions like murder, sadism and rape. Water could also suggest the ephemeral nature of objects in the universe. It emphasizes that everything is transient, like the alternating of joy and sorrow, wealth and poverty. The bath tub occupies a special place in Christianity as the fountainhead of baptism. It is considered in a wider sense to be indicative of the esoteric concept of the endless cycle of births and deaths or reincarnation in Hinduism.

**Illustration 23:** Interpretation of Dreams - Shutterstock _1950003172

# Thirty Nine

# The Solar System

*The world when you look at it, just cannot be random. I mean its so different than the vast emptiness that is everything else, and even all the other planets we have seen, at least in our solar system, none of them even remotely resemble the precious life, giving nature of our own planet.*

**Chris Hadfield (1959 - ) Canadian fighter pilot and retired astronaut.**

Our solar system is an inconspicuous speck in the prodigiously vast universe, planet earth a teeny weeny part of that speck. Yet as per our current understanding planet earth is the only point blessed with a variety of life forms with peak intelligence in the human life form or *Homo Sapiens*.

For Lilliputian man, the solar system is huge, the sun a savage raging inferno with eight planets (excluding the asteroid belt and Pluto) revolving around it that includes our home planet earth. Some lesser known features of our unique solar system:

## MARTIAN BLUEBERRIES:

NASA'S Mars Exploration Rover found *hematite spheres* indicating the possibility of past liquid water and life still not observed and contaminated by humans. Findings fall short of sure fire evidence but could be indicative of past life and possibility of present and future life on Mars.

## JUPITER'S NORTHERN POLAR AURORAE:

These can be described in scientific jargon as *cyclotron driven masers* the only one in our solar system. This is not self explanatory; explanatory notes to assist the reader are appended below:

A *cyclotron* is a type of particle accelerator that accelerates charged particles (initially used in the Large Hadron Collider at CERN Geneva). They have been replaced by vastly superior

## Synchrotrons.

*Maser* stands for *microwave amplification by stimulated emission of radiation*. Its more popular cousin *Laser* stands for *light amplification by stimulated emission of radiation*.

In our present context it is enough to understand that the laser has a wider usage scope than the maser. What is generally not known is that the maser is a *precursor* to the laser.

So what is special about the maser?

Jupiter is the only planet in our SS that *pulses* every 27 minutes manifesting as a burst of beautiful blue light. The discovery made 40 years back, is related to the planet's mysterious auroras and referred to by astronomers as the *northern lights*. Earth's equivalents, the Aurora Borealis and Aurora Australis in the North Pole Arctic and South Pole Antarctic regions pale in comparison to the brilliant display on Jupiter occurring meticulously every 27 minutes. The layman needs to understand the factors causing the phenomenon i.e. planet Jupiter's relatively short day/night cycle of *10 hours*, and a strong magnetic field *20,000 times that of Earth*.

## VENUS THE HOTTEST PLANET:

Although Mercury's orbit is closer to the Sun, yet Venus is hotter.

Why?

The answer lies in the phenomenon called the *Greenhouse Effect*. The greenhouse effect in an earth context refers to energy from the sun getting trapped in the planet's atmosphere, thereby preventing it from returning to space. Earth as a result is much warmer than what it would be without an atmosphere.

When applied to the Mercury Venus duo, Venus has a much thicker atmosphere and retains the sun's heat better. It is like two persons sitting next to a campfire on a cold day, one with a blanket and the other with a sleeping bag. The blanket clad person though more distant from the would retain more heat. If Venus did not have a thick atmosphere its surface would be 128 degrees Fahrenheit colder than 333 degrees Fahrenheit the average temperature of Mercury.

NEPTUNE TRITON & STRONG

## WINDS:

Of the planets in the SS, Neptune is the most distant, the outermost numbering eight from the Sun. It has some special features. Neptune's moon *Triton* surpasses Pluto and *Eris (dwarf planets)* in both mass and size.

A word about dwarf planets. These celestial bodies probably more than 100 in number orbit the Sun but are not accorded the status of regular planets because of their small size and far flung orbits. In respect of Neptune, its orbiting moon Triton is the seventh largest moon in the Solar System.

An unusual feature of Neptune is that it possesses the strongest winds in the SS(1100 mph) that arise out of clouds of varying colours and temperatures. The winds are indicative of rapid changes that take place in the planet's upper atmosphere.

URANUS THE PLANET WITH THE MAXIMUM AXIAL TILT:

Uranus is Neptune's neighbour planet ranked No.7 in respect of distance from the Sun. The planet has an amazing axial tilt of 97 degrees. For comparison, the axial tilt of other major planets are

*Earth 23 degrees,*

*Saturn 27 degrees and*

*Mercury 0.027 degrees.*

For those who are new to astronomy, axial tilt is the angle between the planet's rotational axis and its orbital axis. A planet's orbital axis is perpendicular to the orbital planet or *ecliptic,* the disc surrounding the sun that extends to the edge of the solar system. Uranus is the planet with the maximum change between *solstice* and *equinox.* Most of us are familiar with these terms with respect to planet earth and would be aware that they are caused by the earth's axial tilt.

Solstices designate the point where the Sun's path in the sky is the farthest north or south from the equator. The *summer solstice* occurs around the 20th and 21st of June. The *winter solstice* occurs on the 21st and 22nd December. The summer solstice is the *longest* and the winter solstice the *shortest* day of the year.

The word *solstice* is derived from the Latin word *solstitium* meaning *sun standing still.* It suggests a brief pause as the sun reaches its most extreme point when viewed from Earth before reversing its direction of travel.

The mentioned dates apply to countries in the Northern Hemisphere (likeEurope) and flips over in the countries located in the Southern Hemisphere (like Australia).

There are two equinoxes, the *vernal* on March 21st and the *autumnal* equinox on 23rd September. The While the solstices are characterized by the longest and shortest days of the year, during the equinoxes as the term implies, days and nights are of *equal* duration.

How do solstice and equinox phenomena play out in Uranus?

(Uranus the seventh planet from the Sun is the first planet whose discoverer *William Herschel* an English astronomer is known to us. Mercury, Venus, Mars, Jupiter and Saturn were known to the ancient Greeks. Neptune and Pluto were still to be discovered)

*The next summer solstice in Uranus will be in 2028.*

Why?

Uranus takes 84 earth years to complete a revolution around the Sun. It had a change of season in 2007 in its northern hemisphere in the form of the vernal equinox. The preceding summer solstice took place as far back as 1944. Owing to its extreme axial tilt, there is no change of day and night. For almost 42 years one pole gets constantly illuminated by the sun while the other is immersed in darkness.

The giant planets Jupiter and Saturn that rank fifth and sixth in respect of distance from the sun have much stronger magnetic fields than Earth and more powerful *auroras.* We see only their day side and the auroras remain hidden in reflected sunlight. They can however be studied in the *infra red and ultra violet* ranges of the electromagnetic spectrum.

Saturn's aurora borealis is formed at an altitude of 1200 kms!

Reverting to Uranus, its magnetic poles are nowhere near the geographical ones. The Uranian aurora borealis illuminates the sky just above its polar regions.

## SATURN THE LEAST DENSE BUT SCARIEST PLANET:

Saturn is the immediate neighbour of Uranus with an orbit closer the Sun. The mention of Saturn create images of the spectacular rings that encircle it.

Ever wondered what the planet looks like from close, what comprises those awesome rings?

First things first. Little doubt that the planet looks beautiful from afar. However, like its gigantic gas giant neighbour Jupiter on the other side, Saturn is uninhabitable. You will get killed in a split second if you attempt to set foot on it. The surface temperature is -177 degrees Celsius, bitterly cold.

Saturn is also plagued with storms that could last for months even years. In 2005 there was a thunderstorm on Saturn *1000 times more intense* than any we have seen on earth. One may be tempted to believe that because Saturn is more distant from the Sun with lower gravity compared to Earth it would be pleasurable to *float* on its surface? This is far from being true. The planet's atmospheric pressure is so high (100 times that on earth) that any attempt to float on its surface will crush you to death in a fraction of a second.

Saturn's voice with deafening thunder as recorded by NASA is scary to the extreme. It is in total variance with the heavenly deep sounding *Om* of Earth.

Reverting to Saturn's elaborate ring system, they consist of *icy particles* that orbit the planet. How many rings are there? There is no precise answer. Saturn has more than a dozen rings with in between gaps, the number of gaps being indecipherable as determined by NASA'S spacecraft *Cassini*. Classified as per density, the main rings are referred to as the 'A', 'B', 'C' and possibly 'D' rings the last being the most dense. The total number of rings as mentioned earlier is indecipherable but a calculated guess would be more than 30.

With this introduction on Saturn, it would surprise the reader that it is the *least dense planet in the SS. It is the only planet less dense than water.* With its low density and rapid rotation, Saturn turns out to be the most flattened planet in the SS with an equatorial diameter 10% larger than the polar diameter.

At what stage would the increasing density of a planet earn it the status of a *star?* The answer is when the planet density intensity becomes sufficiently high to trigger off *nuclear fusion* as observed in a typical star like our native Sun.

## OUR PLANET EARTH AND THE MOON:

I have reserved adequate space for our home planet the Earth and its sole satellite the Moon. Have you ever wondered what the Earth looks like from outer space? About a century back this question would have led to considerable speculation; today, space exploration has advanced to an extent where much of the mystery has ceased to exist - a handful of countries USA, Russia, China have sent multiple probes including humans and animals into space beyond the Earth's gravitational pull. Immense data has been collected and analyzed of *the Universe, the Milky Way Galaxy, the Solar System* and our home planet *the Earth* that fits into an *Universal Scheme of things.*

This knowledge frontier is indeed extending at a frenetic pace. Each day seems to reveal unknown secrets of the *macro* universe and its outer external edges. The opposite number the *micro* world and its elusive inner edge seem to defy scientific explanation. The extreme limits in both cases in respective of size, location and time remain *indeterminate.*

Researchers probing these limits are coming around to the view that it is impossible to define *Commencement* and *End* Points. Important as they are, we would like to bypass these profound issues for the time being and focus rather on planet Earth, the only point on our universe that can support the kind of *life* we understand.

There is general unanimity that viewed from space, the Earth is beautiful almost saintly with a rich blue hue caused by 71% water that covers its surface.

Neil Armstrong, the first human to walk on the Moon referred to its beauty by implication when he described the Moon's surface *as an example of magnificent desolation.* Is the Earth the only visibly blue celestial body in the Universe?

No.

There may be a large number of ocean free blue planets *out there.* Indeed there are two in our Solar System, Uranus and Neptune, the two most distant from the Sun. They are *ice giants* with an upper atmosphere containing methane that reflects beautiful shades of *blue wavelengths* of sunlight back into space.

What is the *dark side* of the Moon and

the related question, does everyone see the same side? The time taken for the moon to spin on its axis is *almost* the same as the time it takes to orbit the earth. Hence for an observer anywhere on the Earth's surface, the Moon always keeps the same side pointing his way.

*Over billions of years of exposure to the earth's gravity, the moon is forced to spin synchronously with its orbit.*

*The Moon and the Earth's Ocean Tides:* The brightest and largest object in our night sky, the Moon makes the Earth a more livable planet by moderating our home planet's *wobble* on its axis, leading to a relatively stable climate. It also causes tides, creating a rhythm that has guided humans for countless millennia.

*The Effect of the Moon on Human Behaviour:*

Ever wonder where the word *moonstruck* came from? It stems from the ancient belief that the moon could cause insanity. The first time moonstruck appeared in print is in John Milton's 1674 poem *Paradise Lost* that included it in a list of humanity's afflictions caused by original sin - *moonstruck madness.*

# Forty

# The Vastness Of Intergalactic Space

*Our world hangs like a magnificent jewel in the vastness of space. Everyone of us is part of that jewel. And in the perspective of infinity, our differences are infinitesimal.*

**Fred Rogers (1928 - 2003) American author and Television Host**

This essay should hold you to the edge of your seat in terms of sheer *magnitudes*. *Sizes*, that the human mind thus far in its history and probably forever will be unable to even *imagine* - the vast expanses of intergalactic space.

To qualify for this status, the opening line to the chapter would need to be momentous.

*In the universe, does a billion cubic light-years of* **empty** *space between galaxies contain any mass?*

The calculation that follows shows a surprising result. The mass of this horrendously vast expanse is approximately equal *only* to our humble Solar System i.e. predominately our Sun.

How?

It is estimated that deep space between galaxies contains 0.25 hydrogen atoms for every cubic metre of space. For an earthly comparison, the best man made vacuum contains about a *billion* atoms per cubic metre. Let us probe further:

One light year = $10^{16}$ metres

One cubic light year = $(10^{16})^3$ = $10^{48}$ cubic metres.

A billion cubic light years = $10^{57}$ cubic metres.

Mass of one hydrogen atom =

$1.67 \times 10^{-27}$ kg

Mass of hydrogen atoms in a billion cubic light years =

$0.25 \times 1.67 \times 10^{-27} \times 10^{57} = 4 \times 10^{29}$ kgs

that is about one fifth of the *mass of the Sun*. However, in the calculation we have overlooked two facts:

1. There is 5 times more dark matter than ordinary matter in galaxies and inter galactic space.

2. A billion cubic light years is *nothing* compared to the vastness of intergalactic space. It is normal for galaxies to be separated by *a million light years* so the average void could be about $10^{18}$ (a billion billion) cubic light years or a billion times more than our earlier calculation.

Summarizing, the findings show that despite the incredibly low density of intergalactic space, its mass is comparable to the galaxies themselves a counter intuitive thought, illustrative of the unfathomable and unimaginable size of the universe. We had started the essay on this note and would like to end the same way. However we wish to make a quick review of our reasoning in case something relevant and important has been overlooked. A question that comes to mind is of matter and dark matter in the vast intergalactic space.

Arguing that they should have a presence in intergalactic space identical to the ratio in our galaxy could be cited as an example of erroneous *circular reasoning*. It would seem more logical that intergalactic space would have *less* of both matter and dark matter and related through a different ratio. Another relevant point missed out could be the presence of *rogue stars, solar systems and intergalactic radiation* on a scale where significant mass would be generated but not considered in our calculation.

It seems we have to wait till we know more on the chemical make up of these unknown stars and planets. Speculation without data and laws is *not* science but past experience indicates that intuitive speculation does end up as *law abiding science* in some cases. NASA James Webb Space Telescope (JWST) is a technology front runner that can enable us to move forward. The recent discovery of *Messier 74 or NGC 628* is an example of a hitherto unknown spiral galaxy in the intergalactic expanse. An unusual feature of the galaxy is *the possibility of a black hole at its centre.*

Our next essay, *The Dark Universe* will attempt to probe this abstract topic further.

# Forty One

# A Dark Universe

*Dark matter and dark energy are two things we measure in the universe that are making things happen, and we have no idea what the cause is.*

**Neil deGrasse Tyson American astrophysicist (1958 - )**

I have used the title *The Dark Universe* for this essay in preference to *Our Dark Universe* deliberately. The latter excludes *Multiverses* which is a moot point discussed in scientific circles and does not warrant prime facie exclusion. The essay now presented does *not* rule out the existence of *Multiverses* or parallel universes.

Reverting to *Darkness in the Universe,* for a better perspective we will briefly recount the history that leads to it. In the decade commencing 1990 the scientific view on the expansion of the universe was that its energy density was poised at a level that it would *never* stop expanding but gravity would slow the expansion over aeons of time. This view underwent a change with the advent of the Hubble Space Telescope (HST) that observed that very distant supernovae were in fact accelerating *not* slowing down with time. This observation was counter intuitive and defied explanation based on the known physics of the time. After considering various theories including Einstein's celebrated *Cosmological Constant,* the term *Dark Energy* was coined to explain acceleration at the outer fringes of the known universe as observed by the HST.

We know precious little about Dark Energy. What we do know is that it constitutes about 68% of the universe. We also know that it's equally mysterious cousin *Dark Matter* makes up about 27% and between the two a staggering 95%

of the universe is accounted for. The remaining portion about 5%, is everything observed by us with our instruments on Earth. This is our understanding of the normal *matter* the scientific community talks about.

They have made studies on its behaviour since inception i.e. *birth of the universe about 14 billion years ago.* The rate of expansion *changed noticeably about 7.5 billion years back when objects in the universe begin flying apart at a faster rate.* The concept of Dark Energy was thus born. Einstein tried to explain the phenomenon by defining it as a property of space itself. This challenged the popular concept in an earthly context that space exists merely to *keep or store anything that possesses matter.* Einstein did not face opposition on his theories regarding space and Dark Energy but failed to satisfy the purists. His Cosmological Constant theory held its ground for sometime. It is based on the possibility that that empty space is not *nothing.* It is possible for more space to come into existence. Einstein's CC theory further predicted that *empty space can possess its own energy.* As more space comes into existence, there would be more of this mysterious 'energy of space' that causes the universe to expand faster and faster. This was pure theory and open to scrutiny. An alternative explanation proposed was derived from quantum physics. This proposed the existence of *virtual particles that form and disappear* continuously thereby providing the mysterious energy for expansion.

Efforts to quantify the CC through this theory resulted in the number turning out to be $10^{120}$ *times too big!*

The groping in dark for a plausible scientific explanation for dark energy continued. The concept of *quintessence* proposed by Greek philosophers took centre stage but slowly lost ground due to inadequate back up arguments.

Maybe, Einstein's theory of gravity is incorrect?

Such a shift in theory would affect not only the expansion of the universe but also the behaviour of normal matter in galaxies and clusters of galaxies. The scientific community quickly realized that they faced serious dilemma in theoretical physics from which there was no escape route. Turning to *Dark Matter* after considerable debate we are much more sure *what it is **not** than what it is!*

Firstly it is *dark.* It is not like stars and planets that we can see. There is far too little visible matter in the universe to make up a massive 27%. It is also not composed of *baryons* because baryonic clouds have telltale signs when they absorb radiation passing through them.

Neither are they *antimatter,* because we do not observe *gamma rays* that form when matter and antimatter come in contact and annihilate each other. We can similarly rule them out as *black holes* due to the absence of *gravitational lensing* side effects.

Summarizing, the consensus seems to be that baryonic matter could still make up dark matter if it were all tied up in *brown dwarfs* or in small dense chunks of heavy elements called *Machos* (massive compact halo objects). The other alternative is that dark matter is *non baryonic* but made up of more exotic particles like *axions* or *wimps* (weakly interacting massive particles)

# Forty Two

# Aliens

*The only thing that scares me more than space aliens is the idea there aren't any space aliens. We can't be the best that creation has to offer. I pray we're not all there is. If so, we're in big trouble.*

**Ellen Degeneres (1958 - ) American Television host,
writer and producer.**

The esoteric topic of *Aliens* is often synonymously referred to as the *Fermi Paradox.* For a clear understanding, the reader needs to make a conscious effort to get out of the *earth frame of mind* that confines him to earthly sizes and speeds. The ball game particularly in respect of size is *very very* different.

Our parent Milky Way Galaxy is approximately 50,000 light years in diameter and home to about *100,000 million stars.* Our native Sun is just *a lone inconspicuous* member of this gargantuan medley.

What is really mind boggling is that our universe contains as many as *100 billion galaxies,* and each galaxy in turn almost *1000 billion stars* and over *700 quintillion planets.* I am tempted to play a kid's game by juggling with these mind bending numbers and gleefully announcing that each member of the human race could stake claim to *70 planets in our universe.* Compare these mental capers with the hard realities that have actually occurred on planet Earth in 200,000 years i.e.the period we can stretch human history backwards. We can legitimately feel proud of our achievements, specially the *accelerated rate of growth of technology in the past three centuries.*

We are no longer confined to planet Earth but able to explore our satellite Moon and neighbour Mars. So far our travels have not resulted in an encounter with any alien life but the future brims with exciting possibilities.

It is unpredictable.

Why?

The age of our MWG is 14 billion years, *3 times older* than earth. During this period there could have been many planets where life had evolved. There could have been the existence of *type 3 civilizations.* Where are these people?

Where have the old civilizations disappeared? In an attempt to resolve this quandary, Italian physicist, *Enrico Fermi* has been in interaction with physicists *Edward Teller, Herbert York* and *Emil Konopinski.* The topic covered is *Extraterrestrials and UFO's.*

These could constitute the *type 2 and type 3 civilizations* who could be in possession of technology to travel through interstellar space and the *Dyson sphere* up to the stars. This essentially can only be a game of making *intelligent hypotheses* since it is by its very nature beyond the scope of conventional proof we are accustomed to. Some of the more popular hypotheses are listed below:

- *First born hypothesis:*

  Life in our universe is commonplace and taken for granted. We are admittedly, *the rare intelligent life, the first in 14 billion years with a golden chance to colonize the solar system, the Milky Way Galaxy and there on to other galaxies.*

  The possibility of intelligent species other than *homo sapiens* emerging in the universe in future cannot be ruled out.

- *Brief window hypothesis:*

  Advanced civilizations are situated at a huge distance from us and stopping any interaction. If we send a signal to space at the speed of light it will take the signal *a minimum of 1000 earth years to reach the nearest alien species and 2000 earth years to expect a reply.* Our existing civilization will become extinct and the action continuity disturbed.

- *Aestivation hypothesis:*

This is based on life in the universe having flourished billions of years back *before our solar system and earth* and developed into a highly advanced alien species. Currently they are in a state of hibernation. These early inhabitants were capable of harnessing the energy of the stars and colonizing the sun. They could be waiting for a *perfect condition* to awake from their deep sleep. They are very far away.

- *Rare earth hypothesis:*

According to this school of thought, life as appearing on earth is very rare in the entire universe and restricted to a few planets. These planets have water, energy, and organic molecules and are restricted to a habitable zone of a star. But there are other essential requirements e.g. the correct balance of heat from its parent star and the freezing temperature of space.

- *The zoo hypothesis:*

There are many advanced civilizations in the universe and they are in the process of evaluating the earth site for possible future habitation. According to this hypothesis if life evolved in our universe a few million years back it could have developed into a very big civilization and colonized the entire galaxy. They may be monitoring us and manipulating the lower species. They are waiting for the right time to contact us, maybe waiting for us to reach the correct stage of evolution. As per the review we get the overall picture that life on planet earth is just a microscopic dot in the universe's scheme of things not its *centre of gravity or importance.* It is indeed a humbling thought for earthlings!

What are the views of leading scientific minds on this mind bending topic?

The late British physicist Stephen Hawking had this to say,

*If aliens visit us, the outcome would be much as when Columbus landed in America, which did not turn out well for the Native Americans.*

He was of the view that instead of trying to find and communicate with life in the cosmos, humans would be better off doing everything they can to *avoid* contact. Hawking believed that based on the huge number of planets that are known to exist, we are *not* the only life form in the universe. There are, billions and billions of stars in our MWG alone and reasonable to expect a greater number of planets orbiting them. It is equally reasonable to expect some of the alien life to be intelligent and capable of interstellar communication.

In this context we could ignore the views of Hawking only at our own peril.

There is no unanimity. There are other scientists of eminence who are of a different view who state,

*If they are interested in resources, they have ways of finding rocky planets that do not depend on whether we broadcast are not. They could have found us a billion years ago.*

The biggest and most active hunt for life outside Earth started in 1960, when Frank Drake pointed the Green Bank radio telescope in West Virginia towards the star *Tau Ceti.* He was looking for anomalous radio signals that could have been sent by intelligent life. Eventually, his idea turned into *Seti* standing for *Search for Extra Terrestrial Intelligence* that used the downtime on radar telescopes around the world to scour the sky for signals. For 50 years however, the sky has been silent.

Distance has been a problem in hunting for aliens. If our nearest neighbours are life forms on the forest moon of *Endor* 1000 light years away, it would take a millennium for us to receive any message they might send. If the Endorians were watching us, the light reaching them from Earth at this very moment would show them our planet as it was 1000 years ago; in Europe that means *knights and castles* and in the North American continent small bands of natives living on the Great Plains. The lack of a signal from ET has not prevented astronomers and biologists coming up with a whole range of ideas about what aliens might be like. We can only speculate what aliens look like. In the early days of Seti, astronomers looked for biology similar to our own on the tacit assumption that aliens are likely to be like us. In the early days this was as good an assumption as any. However there are examples in our midst that show exceptions to the rule.

*Extremophiles* are species that can survive in places that would quickly kill humans and other *normal* life - forms.

These single celled creatures have been found in boiling hot vents of water thrusting through the ocean floor, or at temperatures well below the freezing point of water.

*The front ends of some creatures that live near deep sea vents are 200 deg C warmer than their back ends.*

In our supreme ego we call these creatures *extremophiles* that presupposes we are normal. There is the interesting question:

*What does a Martian look like?*

If we relate to the boiling vent creature as a possible answer, the point to realize is that *We are at least as extreme compared to them as they are compared to us.* On Earth, life exists in water and on land but, but on a giant gas planet it can exist only high in the atmosphere, trapping nutrients from the air swirling around it. There is also the possibility that *alien brains* with a different architecture would interpret information *differently from ours* e.g. what we think of as beautiful or friendly might come across as violent to them, or vice versa.

On balance it would be inappropriate to associate human values with aliens because of *basic chalk and cheese* differences. There are several emerging conundrums and the next few decades may enable us to find some of the answers. We need to salute the astro biologists working on the fringes of science and technology to find answers.

Scientists discovered the first few extra solar planets in the early 1990s and, ever since, the numbers have shot up.

Today scientists know of 443 planets orbiting more than 350 stars. Most are gas giants of the Jupiter type, the smallest being *Gliese 581* having a mass of 1.9 Earths. Future generations of ground based telescopes like the

*European Extremely Large Telescope with a 30m main mirror* if operational by 2030 would have enough power to image the atmospheres of faraway planets. It would look out for chemical signatures that could indicate life.

*The Allen Array Radio Dishes,* a gift from Microsoft has at present 42 radio antennae, each 6 metres in diameter. There are plans to increase this number to 300 radio dishes. With the enhanced capability the viewing could increase to 1000 star systems in a few years.

Researchers are optimistic to stumble on an extra terrestrial signal in the next two decades when another million star systems would have come within purview. How should the human community react on picking up a signal?

It would be wise to keep the coordinates of transmitting entity confidential till a full comprehensive evaluation of the source is completed. We do not want random exchange of messages by all and sundry with the all important visitor in our space. There are however humongous difficulties in maintaining secrecy of this data, and leaks are just a matter of time.

# Forty Three

# Origin Of Life

*The complexity of the simplest known type of cell is so great that it is impossible to accept that such an object could have been thrown together suddenly by some kind of freakish, vastly improbable, event. Such an occurrence would be indistinguishable from a miracle.*

**Michael Denton (1943 -) British Australian author, a proponent of intelligent design.**

It is strange that in the midst of incredible achievements in science and technology in the 21ˢᵗ century, humans are unable to answer a fundamental question:

*How did the first life emerge on Earth?*

A recent study of carbon rich meteorites suggests that the building blocks arrived from space. This finding is from Hokkaido University, Japan who detected organic compounds related to the fountainhead of life *DNA and RNA*.

Specifically, researchers investigated, the *Murchison meteorite* that landed in Australia in 1969, the *Murray meteorite* in Kentucky, USA in 1950, and the *fall* detected in Tagish Lake, British Columbia, Canada in 2000. Carbon rich meteorites is a good place to search for clues to resolve this conundrum. DNA and RNA tend to stack together to form nucleo bases of genetic information.

Researchers look out specifically for *pyrimidines* and *purines* in the meteorites. They have *not* been disappointed. The Murchison extracts confirmed what they had long suspected. It led to the general belief that such classes are *ubiquitously present in extra terrestrial environments both within the Solar System and outside it.* Why are these compounds so important?

Strands of DNA and RNA have a spinal cord made up of a *sugar phosphate chain.* Nucleo bases attach themselves to these sugars. In DNA they form the rungs in a helix shaped ladder.

Purine and pyrimidine nucleo bases *always* bind together within DNA due to their structure and the types of hydrogen bonds they can form. A corollary is that the ratio of purine and pyrimidine nucleo bases is always constant within the DNA molecule.

It is important to fit events into the right cosmic time scale - *these nucleo bases, probably the result of photochemical reactions caused by space debris existing **prior** to the formation of our Solar System.* Researchers suggest that during the late heavy bombardment period of planet Earth, 4 billion years ago, the building blocks reached us through meteorite showers.

Samples from asteroids *Ryugu* and *Bennu* are currently under study to establish whether the building block molecules could have a meteorite origin.

What are the views of great scientific minds on this topic?

*Brian Greene,* Director of the Centre for Theoretical Physics at Columbia University, New York has this to say in his book, *Until the End of Time:*

*Could life be such a long shot possibility that it arose only once... Or is life the natural outcome, perhaps even the inevitable outcome?*

We can be sure of one thing only.

Everything we know of now will *cease to exist* over stupendous swathes of time being subject to the inexorable *Law of Cosmic Deterioration.* The breakdown of stars, planets and galaxies is billions of years away. Greene encourages us to think of unfathomable time scales and of the universal forces at work. It is indeed an *intriguing* thought. Can life defy and defeat cosmic deterioration?

These thoughts echo throughout Greene's book and reflected in his comments on *entropy* and *gravity.* But he is not the victim of a single track mind. He analyses *consciousness and intelligence to the trillions of particles that constitute the human form and everything that exists.*

*We are playthings buffeted around by the dispassionate rules of the cosmos.*

*There is **no free will.***

This dramatic conclusion sets up the inevitable interface with *God, Religion and Belief Systems* that have been a characteristic of the human race or *homo sapiens* from the earliest of times.

The unanswered question vis a vis science is:

*Clash or Compromise?*

Before formulating a stand on such a fundamental issue, consider the pertinent factors below:

- Astronomers from Europe used the *Rosetta* spacecraft to show that phosphorus monoxide is a key factor in the origin of life conundrum.

- Data from 66 radio telescopes in the Atacama desert of Northern Chile was used to focus on star forming cosmic area *AFGL 5142* and pinpoint phosphorus bearing molecules in that region.

- The study indicates that new stars and planetary systems originate in such areas making them the ideal spots to prospect for life's building blocks.

*The conclusion is that that phosphorus monoxide, the basic building block of life is found in abundance in identified cavity walls in space that get transported to far away planets like Earth.*

*By Accident or Design?*

# Forty Four

# Religion The Great Dilemma

*I regard the brain as a computer which will stop working when its components fail. There is no heaven or afterlife for broken down computers; that is a fairy story for people afraid of the dark.*

**Stephen Hawking (1942 - 2018) British theoretical physicist and cosmologist.**

This topic in its broadest sense relates to the past history and future prospects of *all* life forms in the universe. It is mind boggling to attempt to fathom what each and every aspect of it could possibly mean. Detailed mental probe shows why this is impossible and perhaps unnecessary.

On a tiny inconspicuous dot in one of the countless possible universes there exists an organism, the *human life form*, undaunted, who with an enquiring mind wishes to squarely face the challenge and uncover as many secrets as possible.

In this context, the words of British philosopher, Bertrand Russell in his book *ABC of Relativity* ring true:

*We know very little, and yet it is astonishing that we know so much, and still more astonishing that so little knowledge can give us so much power.*

This quest is indeed a major project. The first hurdle is *where to start* and thereafter *to find a pathway that could lead to a logical end.* There are a few promising methodologies. I have chosen to follow British anthropologist, *Robin*

*Ian MacDonald Dunbar* (1947 - ) *FBA FRAI* an evolutionary psychologist and specialist in primate behaviour.

He is currently head of the Social and Evolutionary Neuroscience Research Group in the Department of Experimental Psychology at the University of Oxford. He has dissected the topic into three fundamental questions:

1. *The functions that religion has served*

2. *The psychological and neurobiological mechanisms that make this possible.*

   AND

3. *The timing of the origins of religion.*

Dunbar elaborates his theory in narrative form. At the heart of religion, there is *the mystical stance,* that includes a susceptibility to enter trance like states, a belief system in a spirit world that inter alia includes our ability to call on an omnipotent power to help us. As Emeritus professor of Evolutionary Psychology at Oxford, Dunbar's analysis is firmly grounded on Evolutionary Theory. He points out that dependence on religious beliefs entails costs in self imposed pain, celibacy and self sacrifice.

Dunbar thoroughly surveys the evolutionary explanations for religions, the evidence that they makes religious people healthier and happier and their role in building social cohesion. There is historical proof that religious communities tend to be larger, spread faster and last longer than comparable secular groupings.

Dunbar is famous for the number 150. The number in his view is the *natural* size for human groups, the optimum size of a social cohesive group that simultaneously permits good quality one on one interactions with individual members. Dunbar has drawn data from wedding guest lists and Facebook friend lists to support his stand. He also quotes from the experience of hunter gatherer communities of modern societies in which humans have spent more than 95% of their existence. Unlike other primates, humans rely on larger groups for laughing, singing, dancing and feasting. Strangely, this trait in humans lay the first building blocks of elaborate law like forms of religion to keep the social structures intact and prevent breakdowns. It also explains why the great world religions emerged within the very narrow latitudinal band of the northern subtropical zone immediately above the tropics. Deeper study reveals that the ancestral religions were designed to bond very small hunter gatherer communities of 100 - 200 individuals living in dispersed camps of 35 - 50. Dunbar further explains that this served as the prelude to more formal styles of religion incorporating churches, mosques and temples, followed by the emergence of the powerful priestly communities and complex rituals. The initial religion building block groups grew *exponentially* in size and influence. If Dunbar is right in his theory of religion evolution, an inescapable conclusion is that it did not exist prior to 50,000 years i.e. the emergence of the Neanderthal man that *sets humans apart.* The implications of this conclusion are awesome.

How?

*The starting point of religion goes back 50,000 years. Prior to this there was no GOD. The Big Bang Theory estimate of the starting point of the universe is about 14 billion years.*

*Besides, the Dunbar Construct has no explanation for miracles that is an important constituent of major religious belief systems.*

A related issue that crops up is the science/religion viewpoint clash. The Dunbar analysis calls for a review. The death of one of the world's leading cosmologists Stephen Hawking is a good place to commence a discussion on the validity of religion. These are the actual words of the stalwart:

*Humans are just an advanced breed of monkeys on a minor planet of a very average star.*

There have been efforts though to reconcile differences at the highest level.

Pope John Paul II praised Einstein's eminent contribution to the progress of science. Scientism is the theory that the only proper way to answer questions including religious questions is to employ the methods of natural science.

Despite the efforts of atheist scientists like *Richard Dawkins* science has not made religion intellectually disreputable.

An indication of this is to be seen in the appointment of a former Archbishop of Canterbury *Rowan Williams* as the Master of Hawking's own college. In a simultaneous move to confirm the trend an Anglican priest was made the Master of another Hawking college, *Trinity Hall.*

*Francis Collins* who leads the Human Genome project is a Christian by faith.

There are other instances of science/religion camaraderie. Theologian *David Gosling* in a lecture at St.Stephen's College, Delhi said,

"Most of the galaxy of brilliant Hindu scientists of the 19th and early 20th centuries were struck by similarities between certain scientific ideas that resembled ancient Hindu notions. "

At this stage it is pertinent to review contradictions in Christianity having the largest number of adherents in the world today. The idea is not to generate a controversy but to enlighten why *no* religion can be considered *infallible*.

Example 1

*The Sabbath Day*

"Remember the Sabbath day, to keep it holy"

(*Exodus 20: 8*)

in variance with,

"One man esteemeth one day above another; another esteemeth every day alike. Let every man be fully persuaded in his own mind"

(*Romans 14: 5*)

Example 2

*Seeing God*

"I have seen God face to face, and my life is preserved"

(*Genesis 32: 30*)

versus

"No man hath seen God at any time" (*John 1: 18*)

Aside from this, it is relevant to highlight the concepts of *Purgatory* and the God of *Spinoza*. The latter is the preferred alternative for many members in the scientific community to the *Personal God* concept propagated by the monotheistic religions.

*Purgatory*

The earliest mention of purgatory links it to *Hades* in Greece circa 650 - 480 B.C.E. In Homer's *Odyssey*, *Hades* is described as a dark, cold, and joyless place.

A plethora of narratives in the Classical and Hellenistic periods of Ancient Greek history provide graphic descriptions of this intermediate location for those who merit neither a *hero's reward nor a scoundrel's punishment.*

In later Iranian and Zoroastrian eschatology, purgatory is a halfway house between heaven (eternal bliss) and hell (eternal suffering). A stint in purgatory would decide whether an inmate would pass on to heaven in bliss, fall into hell to be tortured or if found to be of neutral merit remain in a limbo like state. Researchers draw a parallel of purgatory with Mount Etna, the stratovolcano in Italy. However the dividing boundary line between purgatory and hell gets *blurred.*

For centuries, Etna's spewing of lava and consequent suffering in the area led to the conclusion that whilst it destroys and damns (hell) it also purifies (purgatory), the latter enabling the inmate to qualify for heavenly bliss. In the event of failure he would find himself in the fiery depths of *Sheol* of the Hebraic bible signifying torture and suffering.

We have provided an oversimplified version of the complex web of Jewish and Zoroastrian belief systems.

### The God of Spinoza

Einstein appreciated the complementarity between science and religion. They were two sides of the same coin. His views on God are summed up in the statement - *What I see in nature is a grand design that we can comprehend only imperfectly...conviction of the presence of a superior reasoning power, which is revealed in the incomprehensible universe.*

He did not subscribe to the idea of a *Personal God* but conformed to the description by Dutch philosopher Baruch Spinoza who had stated - *God is not the creator of the world, but the world is part of God* which is the concept called *pantheism.* It is the antithesis of Jewish and Christian teachings.

He did not believe in free will. His words are - *Free will is inconsistent with the nature of God and with the laws to which human actions are subject. There is nothing that is really contingent. Contingency, free determination, disorder, chance, are products of our ignorance.*

Einstein had opined that although stories from spiritual archives are often dismissed by the scientific community as *myths*, they are not necessarily lies. Scientific purists need to realize that such an inflexible stand could in a broad perspective lead to fatal errors. This duality is brought out beautifully by Sri Lankan Tamil philosopher *Ananda Kumar Swamy's* words:

*Real conflict between science and religion is impossible. The conflicts are always of certain scientists ignorant of spiritual philosophy and fundamentalists who maintain that their mythical story or parable is a historical truth.*

Summarizing, the spiritual reality enshrined in religious myths is by and large lost to the people. The common experience of God that lies at the heart of the mysticism of all religions was experienced by religious mystics. Historically religion has provided individuals with a moral code to live by, a purpose to live for, and a society to live within. Those who welcome the loss of religions influence on public life might recall the woeful history of four nations that banished God:

*Revolutionary France, Nazi Germany, Stalin's Russia and Mao's China.*

There are the fascinating states of *pre birth* and *post death* that has befuddled the human species for time immemorial.

Are the two states identical?

How do they link together providing a continuity for the individual soul?

This query has an interface with major religions of the world.

The monotheistic faiths - Judaism, Christianity and Islam consider it to be *a single event* for an individual with a judgement day for determination of reward or punishment.

The Asian esoteric faiths of Buddhism, Jainism, Hinduism and similar belief systems on the other hand propose an everlasting cycle of births and deaths for an individual where the consequences of his *samsara* and *karma* are played out.

The Asian faiths have slightly different takes on the concepts of *rebirth* and *reincarnation.*

All life forms other than human like avian, marine or land prefer to follow gregarious patterns of flocks, herds, shoals etc.

Reverting to humans it is highly unlikely if not impossible for a single belief system to cover entire humankind. In modern times the *Atheists and Agnostics* also called the *Unaffiliated Group* has grown steadily. They are scattered across the globe and consider themselves as highly evolved individuals. Their numbers presently trail Christianity, Islam and Hinduism but are fast catching up.

The group takes a revolutionary view that a century down the road the numbers professing adherence to today's major religions will drastically reduce and the the *Peoples Republic of China* and/or *Russian Republic* style of living will prevail over the entire globe. This according to them is the inescapable evolutionary trend.

I would like to end this abstract essay on a lighter vein. As a school going kid I was fond of reading comics and my all time favourite was *Captain Marvel*. I remember the sighs of disbelief when the young Billy Batson changed to his *alter ego* the incredible Captain Marvel on uttering the magic word *SHAZAM:*

SOLOMON…Wisdom…*Omniscient*

HERCULES…Strength…*Omnipotent*

ATLAS…Stamina………..*Omnipotent*

ZEUS…Power……………*Omnipotent*

ACHILLES…Courage….. *Omnipotent*

MERCURY…Speed……..*Omnipresent*

Kids of my generation were supremely impressed by the *miraculous materialization* of Captain Marvel and his *superlative* qualities.

The Captain Marvel construct was Godlike if not God himself.

Strangely, this obsession was not at the expense of deities displayed in the family prayer room.

*An example of science / religion coexistence?*

*Is child father of man?*

# Forty Five

# Three Body Problem

*Is it possible that the relationship between humanity and evil is similar to the relationship between the ocean and an iceberg floating on its surface?*

*Both the ocean and the iceberg are made of the same material. That the iceberg seems separate is only because it is in a different form. In reality, it is but part of the vast ocean......*

**Liu Cixin (1963 -) Chinese Science Fiction Writer**

In a general sense, viewed as a standalone expression the *Three Body Problem* relates to the physics of space travel that fall within the domain of the astro physicist. This covers aspects like why a 2 body problem is easier to resolve than a 3 body problem although in recent times frontline scientists have claimed to have found a solution for the latter.

In the 21$^{st}$ century, the title has acquired a political connotation due to Chinese Science Fiction Writer through his trilogy of books. The first is *Remembrance of Earth's Past* followed by *The Dark Forest* and *Death's End*. Taken together the trilogy is referred to as the *Three Body Problem*.

It describes the rise of China in the 20$^{th}$ and 21$^{st}$ centuries from poverty and backwardness to a position of great strength and power to dominate the world. This is not surprising, since the Chinese *Han* race has always believed that located as they are geographically in the middle of the world they are the *Middle Kingdom* destined to provide leadership to the whole world. Although such a viewpoint is debatable, events in the second half of the 21st century relating to the rapid industrialization of the Peoples Republic of China have boosted the credibility factor.

It started as a *pipe dream* in 1950 on the Chinese mainland to rebuild China under their Revolutionary President *Mao Tse Tung*. Liu Cixin starts the trilogy with Chinese scientists struggling to progress in 1970s. This is followed by an imaginary twist in the tale. Humanity makes its first contact with an alien civilization and discovers that not only is the universe full of life but that it is a ruthless battlefield. As the Earth strives to cope and catch up with an unfriendly cosmos, it bears resemblance to 19th century China, a vulnerable kingdom being subjected to incursive bullying by aggressive international powers.

In the novel, the first century of panic after the discovery of alien invaders is marked by political and economic weakness and worldwide infighting, which the author calls the *Crisis Period*.

Humanity's internal turmoil in the novels resonates with themes in the closing stages of the Qing dynasty i.e. end of 19th century /first half of 20th century.

Historically, this was China's *Century of Humiliation* characterized by self imposed isolationism forcibly ended by a rude awakening into the global system.

A system underlined by the Western great powers competing with each other to dominate the world.

Even as the external threat grew, it was internal divisions that drove the deepest rifts in Chinese society in the form of the Taiping and Boxer Rebellions.

The Three Body Problem puts forward a realistic view of symbolic *inter planetary relations* called the Dark Forest Theory arising out of civilization's urge for constant expansion to capture limited resources and expand its frontiers.

The guiding factors are self preservation and elimination of potential rivals.

Communication between geographically distant civilizations breaks down in mistrust and inevitable conflict. Alliances are vulnerable and mendacity as the cornerstone of statecraft rules the day.

The focus again shifts to symbolic anarchic planetary civilizations in the universe that take a lead from happenings on Earth to adopt one of two strategies:

Annihilate a potential rival, or retreat into a shell of self imposed isolationism.

After discovering Earth is not alone, humanity quickly realizes in Liu Cixin's second book, the Dark Forest that the universe is filled with frightened hunters; any sound or flicker of light is a potential threat, so options are to either attack or hide. China's long history is remarkable for its insularism shaped by infighting and attempts to consolidate power. Although more recent scholarship notes the international outlook of the Tang Dynasty and other regional influences such as Buddhism, the primary historical narrative revolves around vacillations between a unified Chinese civilization (Qing and Ming dynasties) and periods of internal political dissensions like the *Three Kingdoms Era* and the *Warlord Period*. Foreign influence on China is recognized as an unfriendly external force acting to foment political destabilization. When faced with external threat or competition, China has tended towards a defensive rather than an offensive riposte. Chinese historical narratives stress the importance of the government's ability to secure itself from foreign incursion, rather than positive references to its previous considerable land conquest. The building of the Great Wall under the Qing and its expansion under the Ming dynasty are good examples of this trend.

In The Three Body Problem, Liu Cixin leaves hope that competition for limited resources on Earth might give way to collaboration for mutual benefit rather than mutual destruction. This concept is conveyed weakly and ambiguously totally lacking in conviction.

A welcome change occurred in China's political renaissance in 1954 with the enunciation of the *Five Principles of Peaceful Coexistence* by China, India, and Myanmar with contiguous borders. Differences in perception on border location and alignment would be settled in due course in a spirit of mutual accommodation. This is the only way forward in respect of problems left behind following the end of European colonial history in Asia, particularly in China and the Indian subcontinent.

This would be in stark contrast to the *Cold War* post World War II between the European powers and USA on the one hand and the Soviet Union on the other.

Liu Cixin's second book the Dark Forest highlights this rivalry. In 2014, President Xi Jinping added the six updates of *sovereign equality, common security, common development, win - win cooperation, inclusiveness and mutual learning, fairness and justice.*

The big question that arises is - Can different civilizations actually work together, or does nature tend towards conflict and destruction?

Chinese leaders have often been accused by Western powers of ambiguous communication and unclear intent. In his novels Liu Cixin takes the view that the only way a civilization can survive is by the masking of real intentions and long term territorial ambitions through continued prevarication. China therefore hedges its bets when formulating its worldview.

Speaking hypothetically, if the emerging 21$^{st}$ century world order results in a return to the Dark Forest reality of anarchic competition, China is capable of walling itself off from foreign threats. If however, principled norms can guide a multi polar world system, China will follow peaceful coexistence. By following such a pragmatic policy, the present Peoples Republic of China is either deliberately or otherwise following the guideline provided by Liu Cixin.

The result is that after a shaky start bereft with teething troubles, the *Great Leap Forward* movement started by Mao Tse Tung based on *To be Rich is Glorious* strategy succeeded in establishing a strong manufacturing base for a wide range of products from *safety pins to jet bombers*. The ongoing forward movement is elevating the status of the Peoples Republic of China in the 21$^{st}$ century to a leading global power.

Aside from the historical work of Liu Cixin that we have outlined, it is fascinating to simultaneously track the pure *physics* meaning of the Three Body Problem. We can do no better than echo the thoughts of the inimitable American astrophysicist *Neil deGrasse Tyson*.

We start with a few definitions related to space exploration i.e. the launching of *manned* and *unmanned* vehicles into space:

*Libration Points* or *Lagrange Points*.

When a satellite is launched from Earth it is held back by the planet's gravity and can escape from it only when its acceleration exceeds *7 miles per second per second*. The satellite would thereafter come under the gravitational influence of the nearest heavenly body the Moon. If gravity had been the only force to be reckoned, then this spot would be the only place in the Earth Moon system impacting the motion of the satellite. But the Earth/ Moon system itself orbits the gargantuan Sun based on a centre of gravity, about a thousand miles beneath the Earth's surface along an imaginary line that connects the centres of the Earth and the Moon. This generates a third force the *centrifugal force* that impacts the satellite, pushing outward, away from the centre of rotation.

From a layperson's viewpoint what is this centrifugal force and do we experience it in our day to day lives?

*We most certainly do.*

Just recap what your body feels like when you make a sharp turn in your car; or when you are making a large circle on a giant wheel in an amusement park?

As the speed of rotation increases you feel the force binding your body parts to the surface getting stronger and stronger. You get the feeling of being *stuck to your seat!*

Eighteenth century French mathematician *Joseph Louis Lagrange* identified spots in the rotating Earth - Moon system where the gravity of Earth, the gravity of the Moon and the centrifugal forces of the rotating system balance.

*There are five such Lagrange Points:*

The first point *L1* falls between Earth and the Moon, slightly closer to Earth than the point of pure gravitational balance.

This point is *precariously unstable* because of the configuration that has given rise to it. The second and third Lagrangian points *L2* and *L3* also lie on the Earth - Moon line, L2 on the far side of the Moon and L3 diametrically opposite way beyond Earth.

*L4* and *L5* are the premium Lagrange points that are *stable.* Their configuration can best be described as being in a valley surrounded by *protectionist hills.* No matter which direction you drift, the forces described prevent you from drifting further.

The excitement in space explorers is that L4 and L5 are ideal sites to position *space stations for research in space and location/communication services on Earth like the GPS.* Today, objects in low Earth orbit like the Hubble Space Telescope and the International Space Station located on L4 and L5 Lagrange Points take just 90 minutes to circle the Earth. (They can be visualized as *apex vertices* on either side of an imaginary equilateral triangle in the heavens with the Earth and Moon serving as the 2 other vertices).

A little thought would reveal that similar to the Earth - Moon configuration there are L4 and L5 points for the Earth - Sun configuration, Mars - Sun, Asteroid - Sun, Jupiter - Sun ..... the list is endless if extended to the cosmos beyond the Solar System. This seems to open up enormous possibilities for space researchers and considering the mysteries of vast swathes of space further investigations are likely to be perpetually ongoing. Recent developments include the commissioning of the *James Webb Space Telescope (JWST)* in 2021 with an expected life of 20 years. It is based on infrared astronomy and has effectively super ceded the Hubble Space Telescope as humankind's Space Observation Tower.

The JWST would probe *formation of the first galaxies* besides exploring *potentially habitable exoplanets in the Goldilocks Zone.*

JWST is currently at its observation point L2 *nearly one million miles away* from Earth.

We shall now continue *to play the game* started by Liu Cixin of alternating between exploration of the space frontier and linking it to the history of the Peoples Republic of China spanning 7 - 8 decades of the last century through his trilogy magnum opus *The Three Body Problem.*

Chinese sci-fi author *Liu Cixin* used the term "Three Body Problem" as a pun for the title of his *Hugo Award-winning 2015 novel.* As the first Asian novel to win the Hugo award, and get nominated for the *Nebula Award for Best Novel* Liu Cixin's achievement is indeed remarkable

The books were written by Liu Cixin in Chinese (Mandarin) and translated into English by *Ken Liu* and *Martinsen.*

The translations contain footnotes that connect to Chinese history like the *Cultural Revolution*

It revolves around *Ye Winjie* a female astrophysics graduate from *Tsinghua University.* Her life sketch consists of a series of challenges ranging from loyalty to her country to the threat of an invasion of Earth by an advanced alien civilization.

The situation is complicated by a collusion interface with a US agency *Trisolaris* planning an invasion of Earth with inter planetary help which due to the vastness of space would take 450 years to reach us. There are many twists and turns in the story, the bottom line being a heady cocktail of treachery, mendacity and betrayal. Interesting as the narrative is, it would not be purposeful to spell out the details since they do not contribute to the message our essay seeks to convey.

Summarizing, the trilogy of books, *Remembrance of Earth, The Dark Forest* and *Death's End* could well qualify as the Chinese riposte to English Science Fiction writer Arthur C Clark's *A Space Odyssey* and futuristic Space travel extravaganzas from the US like *Star Trek* and *Star Wars*.

Though not explicitly stated, China of the 21st century conveys to the science and technology leaders of the West and post World War II Japan that

*In Space Exploration too we are fast catching up. Whatever you can do, we can do better.*

*Three Body* itself is a virtual reality video game in Liu Cixin's narrative. Mind bending concepts abound, the credibility factor getting a boost through periodic appearances of characters resembling *Aristotle, Mozi,* and *Newton.*

Once again, catching up with and exceeding the West is implied.

To ensure that the interest of the plebeian reader in China and more importantly in USA and Europe is not lost Liu Cixin dons the mantle of a story teller. References are made to front line sciences like *pico technology, 11 dimensional super computers,* and use of quantum entanglement by Space invaders to disable Earth made particle accelerators. An objective appraisal of the content by a non Chinese reader would reveal mixed thoughts. Its biggest strength lies in *big sophisticated ideas.*

The novel's *creative concepts* keep the reader in suspended animation and cruise mode till the last page.

An average reader would need more than 6 hours to complete the paper back of over 700 pages. Though highly imaginative and creative bordering on the *fantastic* or even *incredible* discerning readers would place Liu Cixin's trilogy in the top bracket of space narratives that include the possibility of the existence of aliens. It is undeniable however that the author's style of linking of a space fantasy with the history of the China ignoring the rest of the world does tend to confuse the reader in the world beyond China.

This insularity delayed its recognition as an exceptional literary work.

When queried on this aspect, the author in characteristic humility makes repeated appeals to the literary public to consider his effort as *merely a science fiction to entertain the reader - nothing more, nothing less.*

# Forty Six

# The Unknown And The Unknowable

*There are things known and there are things unknown, and in between are the doors of perception.*

**Aldous Leonard Huxley (1894 – 1963)**
**English writer and philosopher.**

This is the last chapter to this book replete with abstract concepts. The *unknown* lies hidden within the folds of the *known*. *Is the world we live in the only one or are there other worlds, invisible, unknown, beyond our imagination?*

It is easy to talk about the unknown but to live it is *surreal.* For starters, can we accept the belief of the monotheistic religions that on an appointed Judgement Day justice will be meted out to the deceased based on their adherence or otherwise to their nominated *Personal God?*

Alternatively, embrace the esoteric concepts of *Samsara, Karma and Moksha* of the religions of Asia primarily Hinduism?

We *do not* have a consensus on this fundamental issue, *probably never will.*

Humans continue to seek answers to questions as old as the hills like,

*Where do we go after we die?*

*Where were we before we were born?*

*Are the above two states identical?*

*Are they two sides of the same coin?*

Similarly a more subtle question that teases humans is whether we have *free will* wired into our system?

Alternatively, are we like non human species and inanimate objects on Earth governed by *determinism?*

Is every action big or small a part of a predetermined scheme?

Is even the rustle of a leaf in the wind on Earth predetermined?

If the answer is *yeah* to even *some* of these questions, the implications for humans on Earth would be far reaching.

Existing norms for *Crime and Punishment, Achievement and Reward* and related norms, practices and conventions considered essential for a decent life would be up for review!

Since the number of *Atheists and Agnostics* or *Unaffiliated* is steadily increasing across the globe, is there a weakening in *Religion* and *God?*

Will science be the sole survivor a hundred years down the road?

Such questions tend to be brushed under the carpet because of our steadily increasing focus on mundane factors like *Roti, Kapda, Makaan* and *Bijli, Sadak, Paani.*

Such thoughts tend to get priority over mysteries of the universe. We can break out of this cocoon of apathy only by adopting a *ruthless* approach to find answers.

*The unknown is the unseen enemy who is also the most fearsome.*

So said *George R.R.Martin* reputed American author of fantasy novels.

We list below a few unresolved issues:

*Incorrect Understanding of Miracles.*

Most people believe that discoveries and inventions are miracles arising out of a *Eureka moment.* This is far from the truth. Viewed holistically, the euphoria that accompanies a successful scientific theory hides years of failed attempts, sweat and toil. Take the case of Albert Einstein. It took him ten long years to formulate the Special Theory of Relativity, and ten more to complete the

General Theory of Relativity. His life long ambition to unify the laws of physics with quantum mechanics remained unfulfilled at the time of his death in 1955. Eight decades later, although there are promising developments within the scientific community like the *String Theory* and *M Theory* we are is still to formulate a Unified Field Theory (UFT) or Theory of Everything (TOE).

In mathematics, *Non Euclidean Geometry* made significant progress only after two millennia i.e. 2000 years!

Alternative geometries were made possible not through lucky flukes but sustained perseverance, through *perspiration* rather than *inspiration*.

Thomas A. Edison, inventor of the light bulb summed up his preparatory efforts in the following words - *I have not failed. I have just found 10,000 ways that won't work.*

*Respect for what came before.*

Students in modern educational institutions across the globe are so used to being served *cooked food in a plate* that they rarely pause to wonder about the arduous journeys and setbacks the creators had to undergo. Centuries of painstaking effort are dismissed with a trite comment or snap of the fingers.

Examples?

The simplicity of the *decimal system* that the modern school student considers trivial took thousands of years to refine and perfect. The words of famous mathematician *Laplace* sum it up beautifully:

*The ingenious method of expressing every possible number using a set of ten symbols (each symbol having a place value and absolute* value) *emerged in India. The idea seems so simple nowadays that it's significance and profound importance is no longer appreciated. Its simplicity lies in the way it facilitated calculation and placed arithmetic foremost amongst useful inventions.*

The decimal system emerged in India in the era of two of the greatest scientific minds of the West - *Archimedes* and *Apollonius!*

Another example that comes to mind is our struggle to accept that our Solar System is *Heliocentric and not Geocentric.* In retrospect it seems highly regrettable almost shameful that *Galileo* (1564 – 1642) who authored *Dialogue Concerning the Two Chief World Systems* that rejected the Geocentric Ptolemaic system faced such vehement opposition from the Church that it was placed on the *Index of Forbidden Books* for over 200 years!

In our present book we have devoted fifteen chapters to *Math Conundrums.*

This is in recognition of the role mathematics has played in extending the knowledge frontier. It is a tool that has enabled humans to understand and predict scenarios in the far reaches of the universe that cannot be done physically. Despite advances in space technology, we are still *babes in the wood* in physically exploring the unknown knowledge frontiers. We are microscopic sized entities on an inconspicuous planet orbiting an average star in one of the billions of galaxies known to exist. We are keen to establish contact with our counterparts, if they exist elsewhere in the universe. Although It does seem we are unique, the prodigious vastness of the universe precludes a precise answer.

It is impossible to deny that Earth like or even superior *life* and *intelligence* could be lurking elsewhere eager to make contact with us. We have to acknowledge significant advances in technology in the 21$^{st}$ century and ongoing progress at a frenetic pace have to be acknowledged. Consider, we have been able to land humans on the Moon a few decades back. Scores of satellites orbit the Earth that facilitate telecommunication services.

Probes have been sent to the far reaches of our Solar System and outer space.

Research findings require validation which *ipso facto* imposes limitations and challenges relating to both *macro* and *micro* worlds.

In this respect mathematics holds a privileged position. It is the unique discipline that enables the human brain to *understand, interpret and predict* knowledge frontiers with a modicum of equipment and instrumentation.

Humans have refined it to an extent that *every event in the universe irrespective of size and time* falls within our ability to probe.

*#The Special Role played by Mathematics.*

The following points from the history of humankind are relevant to our narrative:

- Indians and Chinese developed distinct, and sophisticated, mathematics.

- Sumerians and Babylonians around the same time developed the *sexagesimal* (base 60) *system.*

Because of its non glamorous image compared to other fields of human endeavour mathematics had been forced to take a back seat. Defying this trend were mathematicians from across the globe listed below:

\# *Elizaveta Litvinova* Russian, obtained doctorate from Switzerland in 1878 held in high esteem in Russia as a pedagogue.

\# *Sofya Kovalevskaya* (1850 -1891) also from Russia who despite adverse social norms worked under the famous *Karl Weierstrass* in Heidelberg, Germany. Eventually awarded doctorate by University of Gottingen.

\# *Sophie Germain* (1776 - 1831) French mathematician, an autodidact who mastered works of Newton and Euler. She developed a healthy working relationship with the well known mathematician Gauss.

\# *Julia Robinson,* American mathematician whose words ring true today:

*Mathematicians form a nation without distinctions of geographical origins, race, creed, sex, age, or even time - all dedicated to the most beautiful of the arts and sciences.*

Nazi Germany accorded selective shabby treatment to mathematicians based on ideology and race. *David Hilbert,* the eminent mathematician when questioned about Gottingen, the Mecca of Mathematics at that time stated

*Mathematics in Gottingen?*

*There is really none anymore.*

\# *Cora Ratto de Sadosky* (1912 -1980) was an Argentinian mathematician who spent her life fighting racism. She was forced to flee to Barcelona Spain to escape from the oppressive military regime that had taken over her country.

She died in exile unable to fulfill her cherished dream of returning to her native country.

- *Mathematics is Humanity's Best Shot at Understanding the Universe.*

While we understand only a minuscule part of the totality of information, the days of superstition and irrationality are over. A good example is how eclipses occur. With modern science and mathematics we understand *how* and *why* they occur and are able to *predict* nature and timing of future eclipses of the sun and moon. While it is educative to study the past achievements of mathematics we would emphasize that *it is only a means to expand the knowledge horizon and not an end in itself.*

# Forty Seven

# Conclusion

It is not easy to comprehensively conclude this final chapter on THE UNKNOWN AND THE UNKNOWABLE. We are attempting to provide a summary of our main findings below:

*Our universe contains everything that exists - all of space and time and whatever there is anywhere. More specifically there are approximately 200 billion galaxies. They exist in clusters and super clusters that are scattered and separated by thin galactic filaments. The humongous size of these astronomical objects can scarce be imagined. For example the Hercules - Corona Borealis Great Wall, the largest galactic filament discovered in 2013 is estimated to be 10 billion light years in length.*

*For a better perspective of this prodigious immensity, it is to be noted that our parent Milky Way galaxy that we consider as huge has a diameter of 180,000 light years that is a teeny weeny 0.001% of the HCBGW!*

*Aside from this, there are great voids between clusters and filaments.*

*A remarkable feature of the cosmic web is that it resembles the human brain.*

*Neurons in the human brain too, form clusters, that connect through thin nerve fibres called axons.*

*Does this mean that the universe could have thinking capacities similar to the human brain?*

*Also are there differences?*

*Unlike the human brain that does not move, the universe from the Big Bang onwards has been expanding and accelerating to the present moment through a process that seems to extend to a never ending beyond.*

*The speed of signal transmission in the human brain is yet another important difference. Neurons are capable of transmitting 5 - 50 signals per second.*

*In stark contrast, signals at the speed of light, would take 90 billion years to travel from one side of the gargantuan universe to the other.*

*Should these differences compel us to conclude that the universe is not thinking very much?*

*The concept of locality with reference to the universe is debatable.*

*We do not know why the universe respects locality at a macro level. We also do not know whether locality works in the subatomic micro end of the size spectrum. Despite such uncertainties, the consensus in the scientific community is to favour the existence of a non locally connected universe.*

*Crazy as it may seem, the idea that the universe is intelligent and capable of thinking is compatible with our present day knowledge.*

*What about unresolved issues like*

*Negative Mass?*

*Faster than light communication through the Quantum Internet?*

*Establishing contact with parallel universes?*

*We are unable to provide clear unequivocal answers to the above questions.*

*There is a great deal about the universe that we still do not know!!*

# Selective Bibliography

**Part I The Human Life Form**

**Books**

- 1969 Reader's Digest Asia Ltd Hong Kong

  *Our Human Body Its Wonders & Its Care*

- Readers Digest Library of Modern Knowledge Volumes 1 & 2

  *The Human Body*

  *The Human Mind*

  *Language and Communication*

- The Cell by John Pfeiffer and the Editors of Time Life Books 2$^{nd}$ Edition Hong Kong

  *The Cell - A Bustling Metropolis*

  *The Architect and the Master Builder*

  *The First Living Thing on Earth*

  *The Message Carriers*

- The Body by Alan E Nourse and the Editors of the Time Life Books 2$^{nd}$ Edition Hong Kong

  *Hallmarks of Individual Identities*

  *The Heart and its Couriers*

  *The Vital Pairs of Lungs and Kidneys*

  *Fuelling the Body's Machinery*

*The Productive Power of Hormones*

- Quest for the Unknown – Mind Power

  A Dorling Kindersley Book

  *Mysteries of the Mind*

  *Mind over Mind*

# Websites

Salmon run: Wikipedia **https://en.m.wikipedia.org**

Salmon Reproduction: MarineBio.net

**https://www.marinebio.net**

The Salmon Life Cycle: National Park Service

**https://www.nps.gov>learn**

Human embryonic development: Wikipedia

**https://en.m.wikipedia.org**

Embryo Development - A Development process of Fetus

**https://byjus.com>biology**

Lung Anatomy, Function, and Diagrams

**https://www.healthline.com**

Liver: Anatomy and Functions/John Hopkins Medicine

**https://www.hopkinsmedicine.org**

Heart: Anatomy and Function – Cleveland Clinic

**https://my.clevelandclinic.org>body**

Human Brain Anatomy

Anatomy of the brain

**https://mayfieldclinic.com>...**

Human Kidney Anatomy

Kidney - Wikipedia

**https://en.m.wikipedia.org>...**

Endocrine System Male - Female

Endocrine System Explained in Males And Females/BioIsland

**https://www.bioisland.com**...

The Third Eye - Wikipedia

**https://en.m.wikipedia.org>**...

How to Open Your Third Eye Chakra for Spiritual Awakening

**https://www.healthline.com**...

Tower of Babel - Wikipedia

**https://en.m.wikipedia**...

**Part II Math Conundrums**

**Books**

- Orient Publishing (A division of Vision Books Pvt. Ltd) New Delhi - 110 002,

  1st Published 2010

  *The Little Book of Maths Theorems, Theories & Things by Surendra Verma*

- Coronet Books, Hodder and Stoughton,

  First published in Great Britain 1977

  *Figuring The Joy Of Numbers by Shakuntala Devi*

- *Published by Dell Publishing Co INC

  1961.

  *Of Men And Numbers by Jane Muir*

- First Published 1943, Penguin Books Ltd, Middlesex, England,

  *Mathematician's Delight by W.W.Sawyer*

- Readers Digest Library of Modern Knowledge Volumes 1 & 2

  *Mathematics*

- Mathematics by David Bergamini and the Editors of the Time Life Books - 2nd Edition Hongkong

  *Numbers - A Long Way from One to Zero*

  *The Shapely Thinking of the Ancient Greeks*

  *Figuring the Odds in an Uncertain World*

**Websites**

The Number of the Beast is 666

Number of the beast - Wikipedia

**https://en.m.wikipedia.org>wiki**

The Most Beautiful Equation - Euler's Identity

Wikipedia

**https://en.m.wikipedia.org**

The Symbol Pi ($\pi$)

Pi/Definition,Symbol,Number,&Facts/Britannica

**https://www.britannica.com>science**

The Symbol Phi ($\varphi$)

Phi - Wikipedia

**https://en.m.wikipedia.org**

The Problem of the Grazing Goat

Quanta Magazine

**www.quantamagazine.org**

The Hare and the Tortoise

Byjus

**https://byjus.com>kids-lea**

Carl Friedrich Gauss

Wikipedia

**https://en.m.wikipedia.org**

Carl Friedrich Gauss

Encyclopaedia Britannica

**https://www.britannica.co**

Leonhard Euler

Maths history.st-Andre

**https://mathshistory.st-and**

Maths is Fun

**https://www.mathsisfun.com**

**Part III Knowledge Frontier Areas**

**Books**

- Create Space Independent Publishing Platform North Charleston, South Carolina, USA

  *Our Universe: An Unending Mystery*

  *Raman Sivashankar*

- Bantam Books Trade Paper backs, New York, September 1998

  *A Brief History of Time by Stephen Hawking*

- Paperback edition published in 2006 by University of Notre Dame, Indiana 46556

  *Modern Physics And Ancient Faith by Stephen M.Barr*

- Oxford University Press, London

  Published in India in 1957 by John Brown Oxford University Press, Madras

  *Outline History of Civilization by F.G.Pearce*

- First published in USA by Harmony Books, Crown Publishers Inc in 2000

  *How to know God by Deepak Chopra*

- First published in Great Britain by Vintage Books, London in 1993

  *A History of God by Karen Armstrong*

- First published in India in 2011 by Harper Collins Publishers India

  *The Essential Koran by Thomas Cleary*

- Published by Sri Sathya Sai Books and Publications Trust Last Edition in 2007

  *The Divine Life and Message of Sri Sathya Sai Baba*

- Published in 1988 by Harper & Row, New York

  *Tao Te Ching by Stephen Mitchell*

- Readers Digest Library of Modern Knowledge Volumes 1 & 2

  *Structure of the Universe*

  *The Nature & Origin of Life*

- Quest for the Unknown - A Dorling Kindersley Book

  *The Cosmic Clock*

  *Prophetic Revelations*

  *Traveling in Time*

**Websites**

Worm Hole

Home - Wormhole

**http://www.wormholebridge-to**

Quantum entanglement

Wikipedia

**https://en.m.wikipedia.org**

Interpretation of Dreams

Methods of Dream Interpretation: What Do Dreams Mean?

**https://www.verywellmind.c**

**SHUTTERSTOCK IMAGES 1-23**

SS Images that appear after Part I, Part II and Part III